THE TOTAL
INVENTOR'S
MANUAL

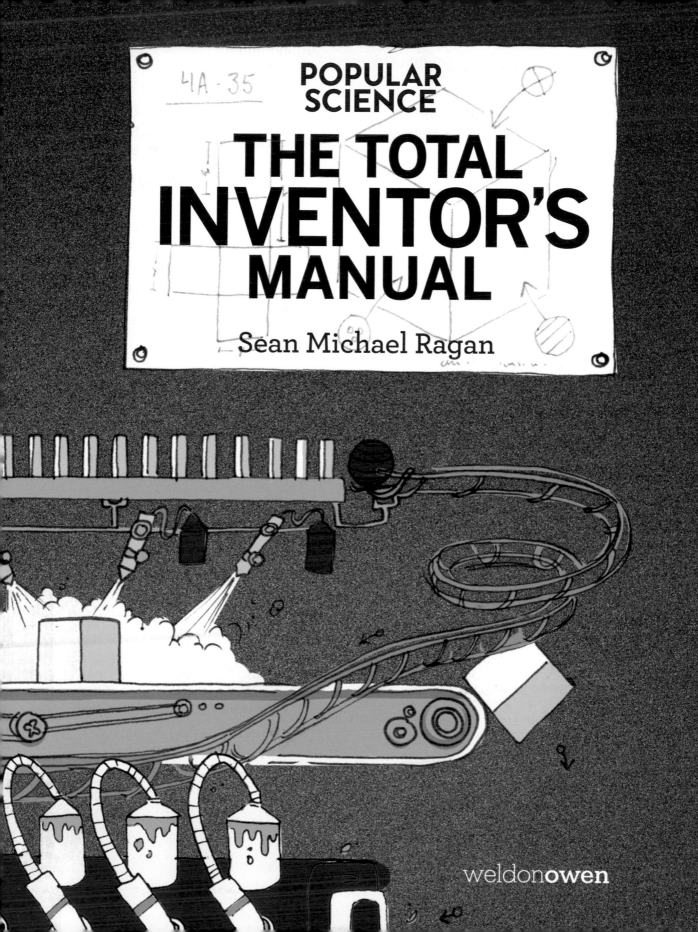

CONTENTS

Introduction

MAKING ONE

ON THE CARE AND FEEDING OF IDEAS

THE PROTOTYPING CYCLE

THE TESTING PROCESS

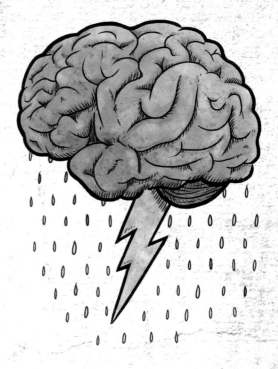

SELLING ONE

FINDING FUNDING

MAKING IT PRETTY

SWIMMING WITH SHARKS

SUPPLY-CHAIN MANAGEMENT

SELLING MANY
MAKE IT TO MARKET

CUSTOMER SUPPORT

SELL OUT...OR SELL ON

INTRODUCTION

This book is dedicated to my father, Lawrence Henry Ragan, who lived the dream. Born into poverty on the windy West Texas plains eight months before Pearl Harbor, he is now 75 years old, with 30 patents to his name, in fields ranging from wristwatch telecommunications to drone aircraft design. He still gets up before dawn every day, heads to the garage, and works on perfecting number 31. So far, he has founded three companies and sold two of them. The third, as of this writing, is almost 40 years old, generates annual revenues of more than US$30 million, and employs nearly 70 U.S. citizens.

Dad taught me to use tools, to scavenge parts, to fix broken things, to do it myself instead of paying someone else. He taught me to take machines apart to see how they work, and for the challenge of putting them back together. He taught me to *tinker*, which is a word in dire need of an overhaul. Look it up, and you'll find suggestions of idle, unproductive activity—of time that could be better spent. And in a narrow economic view, that definition seems sensible: A woman whose hours are worth US$100 apiece is surely a fool to spend two of them repairing an appliance she could replace for US$50, right?

That logic treats the experience (including the understanding she gains) as worthless. But we now live in a world where heroes can be assassinated by robots, tyrants toppled by cell phones, and millions made or lost by clicking a few links on a computer. In such a world, power depends as much upon understanding as it does upon wealth—specifically, understanding the technologies that make all these horrors and wonders possible. Those who believe in the democratic ideal—who see themselves as citizens rather than subjects, entrepreneurs rather than employees, masters rather than slaves—face a choice: Understand how these things work, or be at the mercy of those who do.

While reading, figuring, and formal schooling each has its place in achieving that understanding, none is more important than the humble art of tinkering, the process of teaching oneself with one's own hands and brain. Without this joyful, creative, hands-on experimentation—without *play*, in other words—even the best education hangs lifeless and empty, like a fine suit forgotten in a back corner of the closet. We may pull it out for a job interview or an exam, but it's just a costume, part of a role we play to get along in life.

In my ten years of writing about engineers, makers, artists, inventors, and other "doers of things," I have come to believe that no single practice is as crucial to our future as tinkering. Those who get good at it may invent new and useful things. A small fraction of these may, in turn, go on to start companies and generate wealth. If that's what you want, this book can help: Take it and read it, then go out and do it.

But that's not the most important thing. Not everyone can turn out to be the next Steve Jobs (or even the next Lawrence Henry Ragan), but everyone can learn to use tools, to scavenge parts, to fix what's broken, to take things apart just to see what's inside. Everyone can learn, in short, to tinker. And if this book helps get you there, it will have done its job: You will have a powerful new tool to help navigate the future, and I will have done my bit to pass on the most precious thing I have ever been given.

I love you, Dad. Keep dreaming.

Sean Michael Ragan

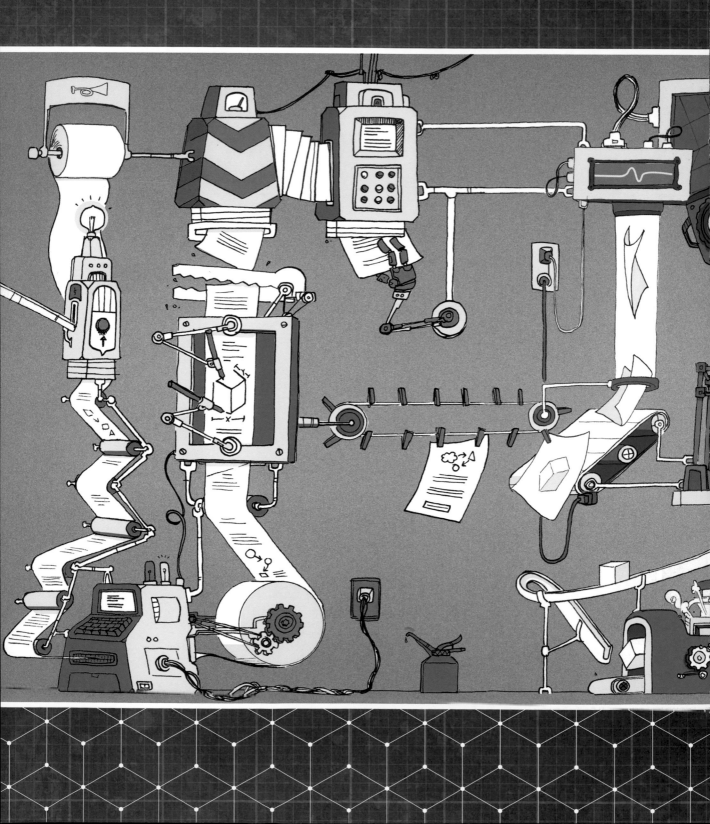

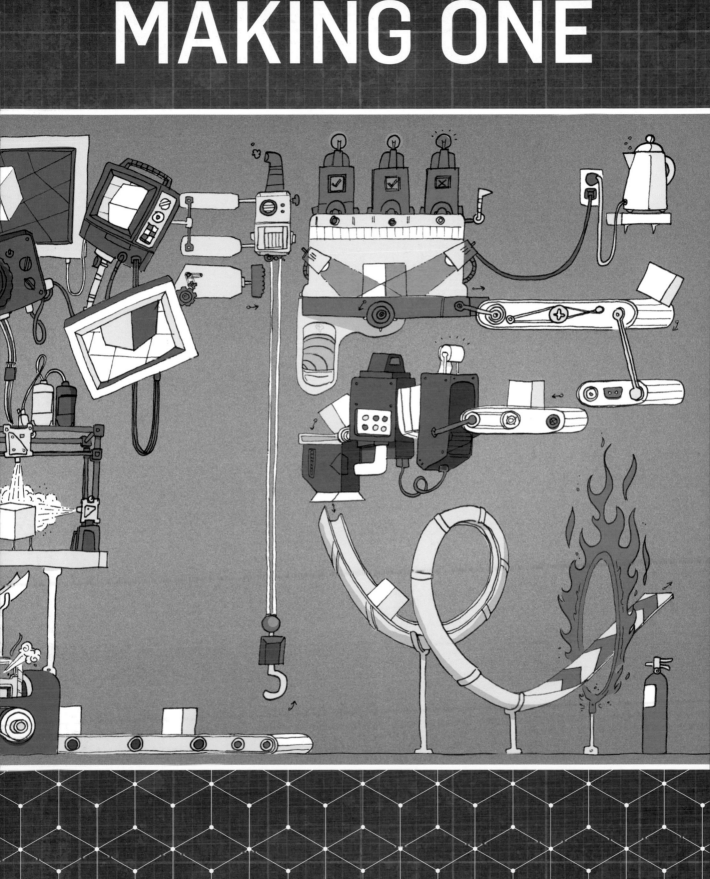

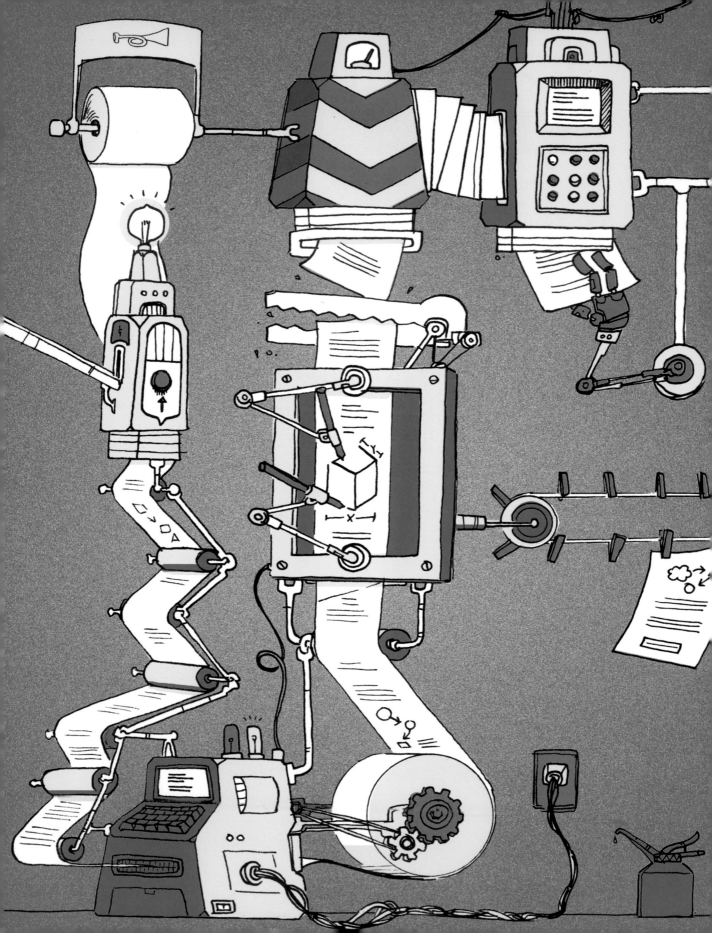

ON THE CARE AND FEEDING OF IDEAS

Like drawing, doing algebra, or speaking a second language, having original ideas is a mental skill that can be developed and, with practice, become second nature. Whether you dream of creating the next big thing in the tech space, need a concept for a pressing assignment, or are just looking to transform your mind into a fertile breeding ground for good ideas, try a few of these practical tips to shake up your sluggish muse.

001 GIVE YOURSELF PERMISSION

Society doesn't always encourage us to be creative. At a very young age, we start learning what's expected to fit in, and our own original ideas start to seem less important. So we get out of the habit of having them. And when, as adults, we're called on to produce creative work, we may be plagued by insecurity or negative self-talk: *I'm not an ideas person. I suppose I might try such-and-so, but that's stupid and would never work.*

First, don't think that way if you can help it. Second, even if you can't silence your inner critic, remember that first ideas are often stupid and almost never work. A genius isn't so much someone who has better ideas than everybody else as someone who just has more of them—and is unafraid to dig through dozens of bad ones to find a diamond in the rough. Anyone can do this, and the more you work at it, the easier it gets.

002 CLEAR SPACE IN YOUR HEAD

In 1991, a bright, young artist named Tom Friedman had his first solo exhibition in New York. A conceptual sculptor, Friedman's works often consist of everyday objects like pencils, soap, and aluminum pie pans arranged in striking ways. He's created explosions out of toothpicks, cadavers out of paper, and giant spheres out of chewed gum. He made one of his early works by mounting a spinning canvas to a wall and signing his name on it over and over as it turned, spiraling toward the center as his pen gradually ran out of ink.

Today an original Friedman easily fetches US$150,000. But before his glory days, he was a grad student at the University of Chicago, where his journey to find the creative potential in everyday things started by closing himself in an empty, all-white room: "Every day I would bring an object from my apartment and place it somewhere in the space. The first day I placed a metronome on the floor, and it just clicked back and

forth. Or I would sit the whole day, on the floor, looking at it and thinking about it, and asking questions about my experience of it . . . For me, this was more like a mental space that had been cleared away."

Your own empty room is likely to be metaphorical—maybe it's meditation, exercise, hot baths, or whatever else works—but the point is the same: Eliminate the distractions of your daily life and listen to what bubbles up in the silence. The longer you listen, the more interesting the things you will hear.

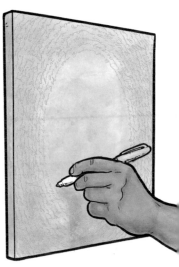

003 WRITE IT DOWN RIGHT AWAY

Ideas are guests in your brain. Be sure to treat them with hospitality. Moreover, they're fickle guests who come from a mysterious place, drop in unannounced, and may not be inclined to stay that long. Who knows how many good ones have been lost over the eons because they chose to call on someone distracted by hunting, eating, sleeping, socializing, or some other more practical obligation?

Don't be that person. When an idea shows up and makes you go "Aha!", stop what you're doing (safety permitting) and immediately make some kind of record. Get it on paper, into a computer, or somewhere on your phone within a few seconds. Snapping a quick picture of whatever triggered your brainstorm is better than nothing, but be sure your process includes filing all these seeds away in a single, easy-to-search location, so you'll know how to find them later. I like to keep a plain text idea file on my computer desktop, and I pay a remote backup service a small annual fee to automatically upload the file to a data vault whenever I add something.

TOOL TIME

004 PROTECT YOUR IDEAS WITH A NOTEBOOK

For decades under U.S. patent law, the inventor's notebook was not just an eccentric prop to go alongside the bubbling chemicals, lab coat, and propeller beanie—it was an important legal document. When a dispute arose about the awarding of a patent, the courts would investigate to determine who had been first to invent the idea, and a properly maintained research notebook was the most important piece of evidence. So the patent rights—and the potentially lucrative legal monopoly that goes along with them—would generally go to the inventor who kept better notes along the way.

In 2013, federal law was changed to award priority to the *first inventor to file* a patent application with the U.S. Patent and Trademark Office, which brought its processes more in line with those long practiced in most of the world. That change has pros and cons, but one upshot is that the importance of the notebook as a legal document has been somewhat diminished.

Regardless of the patent laws in your area, there are still lots of good reasons why you should have a proper notebook (see #015) and be disciplined about keeping it up-to-date. First and foremost, it's a critical aid to fallible human memory. Inventing something worthwhile is a long, laborious process, and the time and energy you invest in figuring stuff out is wasted if you can't later recall what you did and how you did it.

005 PERUSE THE PATENTS

Philosopher George Santayana once said, "Those who cannot remember the past are condemned to repeat it." It's oft quoted in the context of history, but it's just as true with inventing. To bring your work to market, you need to know what's come before.

Start by doing an extra-thorough online search. Think about the many ways someone might describe your idea, and try those searches too. Scientists joke that "a year in the lab will save you a day in the library." Believe it: If you've got a killer idea, you may be disappointed to learn that somebody else has already been there, done that, and sold a million T-shirts. So it's better to learn now than it is to waste time on a dead end.

Next, search the patent files for prior art: published patents and applications that may describe a similar idea (see #106). All the world's major patent databases are freely searchable online, so dive in.

006 HONOR MOTHER NECESSITY

Everybody knows the old saw about necessity being the mother of invention. At the risk of bad taste, there's been a lot of speculation about who donated the Y chromosome. Galileo once wrote that "Doubt is the father of invention," and other thinkers have pointed the finger at a whole cast of downright mythological characters, including opportunity, scarcity, curiosity, laziness, and ingenuity. All we really know about Papa is that he's shiftier and harder to pin down—a bit of a rolling stone.

But if your goal is to invent something profitable, it's usually best to take the conventional route and start by looking for problems wanting solutions. Train yourself to think past the status quo and imagine how the world could be better. Just because we've all been doing the same thing—and in the same way—for as long as anyone can remember doesn't mean it couldn't be done using a better method.

And if you're ambitious, consider Problems Writ Large—be they problems of the economy, the environment, medicine, or the human condition itself. Here you can take your pick of history making examples: the printing press, the polio vaccine, the light bulb, the iPhone. Coming up with a game-changer that improves the lives of others requires paying close attention to the world's troubles, and talking with people on the ground to develop a deep understanding of what may help. Saving the world through invention is not easy; it has big problems, they're hard to solve, and there are lots of smart people trying to solve them already. But if you feel the calling, by all means: Get in there. We need you.

007 START CLOSE TO HOME

While inventing something world-changing is a worthy goal, you shouldn't overlook the more humble problems that are likely lurking right under your nose. Think of all the pesky everyday nuisances that might be reduced or eliminated with a clever new tool, trick, or gizmo, thereby saving yourself and others like you time, money, energy, or stress. Or go after the super-specific problems that enthusiasts only discover when they're deep in the nitty-gritty details of their work. Here are some questions to ask.

WHAT IRRITATES YOU? Or your friends, colleagues, or family members? (Besides each other, of course.) Pay attention to the struggles around you, and you may just spot a trend. How else do you think the salad spinner came into this world?

WHAT TOOL FRUSTRATES YOU? If there's a device that gives you grief (a seat belt, garden shears, or a smartphone case—anything), how could it work better?

WHERE DO YOU WASTE TIME? We all do it. Is there a device that could help you out of a specific time trap?

WHAT SPACES NEED IMPROVEMENT? We spend our lives in environments made by fallible human beings. What would you do differently if you were redesigning your bathroom, kitchen, workplace, computer, or car?

HOW COULD YOUR FUN BE MORE FUN? Don't forget your hobbies. Could your roller skates have more ankle support, or the buttons on your video game controller stick less? Believe it or not, these problems matter to a lot of people. So be a hero and fix one.

008 WORK WITH WHAT YOU'VE GOT

What I call top-down inventing means picking a problem and working to solve it. But there's also the bottom-up approach. Take the familiar Pet Rock. Does it really solve a problem? If you answered yes, I'm prepared to agree that locals running tourist traps have a "problem" in producing a low-cost souvenir that can be sold at high markup to visitors, but I draw the line at "necessity." No one *needs* a Pet Rock, and yet it's an extremely profitable invention. Here the trick is to look for available resources and opportunities. What are you good at? What do you have a lot of on hand right now? How can you add value to it and sell it to people?

010 FIND YOUR TRIBE

At the genetic level, you're a pretty special snowflake. At last count, there were more than 7 billion human beings living on Earth, and while we all share more than 99 percent of our DNA, the remaining fraction of a percent still contains more than enough information to uniquely identify every one of us (except identical twins, of course).

When you consider your needs, desires, hopes, dreams, and problems, however, you become a little less rare. If the Internet has taught us anything, it's that no matter how bizarre your psychology, you're not alone in it—there are likely not just one or two but an entire community of people out there who, if you knew them, would seem eerily similar to yourself in intellect, interests, goals, outlook, and personality. So if you've invented something that you, yourself, find valuable, chances are you're not the only one. Chances are, there's a market. And, again thanks to the Internet, it's now astoundingly easy to find and reach that market. (Tip: Treat these people right, and they may come to your aid throughout the process, helping later on with testing, funding [see #089], and more.)

Admittedly, if you can address a more common problem—like that whole getting old and dying thing, for instance—you're going to have a bigger potential market and bigger potential profits than if you'd invented a new hypoallergenic shoe polish for fetishists. But then, what are your goals? Are you trying to start a business and make a lot of money? Or just help out other people like yourself by providing a clever, useful thing? You may be surprised how the latter can result in the former.

EXERCISE

009

GO ON A "JUNKET"

Practice looking at familiar objects in new ways, and you'll start to see the world as an inventor does. Everything everywhere can be a seed for new ideas.

STEP 1 Reach deep into your junk drawer (you do have a junk drawer, right?) and pull out a random object. Be careful doing this—there may be sharp, rusty, or other ouch-y stuff in there.

STEP 2 Depending on the extent to which you express the hoarding gene, you may be dismayed to find yourself holding an ugly and apparently useless piece of trash, like a bent nail. Even if so, your task is to make something beautiful and/or useful out of it. Give it a good stare: What is its shape? Its intended use? What could be its *unintended* use? Does it have any interesting features that you could exploit? What tools or processes could you use on it? Mechanical? Electrical? Chemical?

STEP 3 Stick with it until you have a satisfying "aha" moment, and make some kind of record of your invention.

011 TAKE A CUE FROM THE CUECAT

This bewildering thing is a bar code scanner in the shape of a cat. If you bought the book you hold in your hands in a store, someone used a similar device to ring you up at the register—this is the same technology, only in cat shape. The cat's "tail" is a cable with a split-off cord that allows you to connect it to your computer's keyboard port and the keyboard at the same time. Thus connected, the CueCat draws power from the computer, just like the keyboard, and sends signals that the machine interprets as keystrokes. Special software on your computer takes care of the rest.

CueCat was launched in 2000, with much fanfare. The idea was that people would use it to interact with physical objects—magazines, advertisements, soda cans, what have you— without having to type anything in, much like the QR codes of today. So you could scan the bar code on your latest issue of *Popular Science* and be taken right to a website where you could learn, and hopefully buy, a lot more.

This happened back in the early days of the Internet, before we all carried it around in our pocket on a smartphone. All people knew about this online stuff was that it was going to be big. Few understood it, and those who did were quick to capitalize. CueCat was funded to the tune of US$185 *million* by a slew of major corporate investors, like Coca-Cola and Radio Shack (remember them?). Of course it flopped— and US$185 million is a lot of shirts to lose.

The thing is, the CueCat was cheap and worked just as intended. In fact, there was a thriving used market for them for a long time: With a scrap of hacked code, you could use an old CueCat to scan just about anything with a bar code on it. What CueCat got wrong, as a brand, was that it grossly overestimated the size of its market. Not many people wanted to interact with their soda cans, and those who did were willing to type in a few keystrokes rather than keep a dedicated widget around for the purpose. It solved a problem nobody really had.

012 DETERMINE IF YOUR IDEA IS A GOOD ONE

So you have an idea. How do you know if it can bring you fame and fortune—or if someone else has already taken a similar notion to market, where it crashed and burned in an eyebrow-singeing fireball visible as far as the eye can see? Follow this handy flowchart to discover if your product is honestly worthy of development—and if you have the resources, mindset, and commitment to see it through.

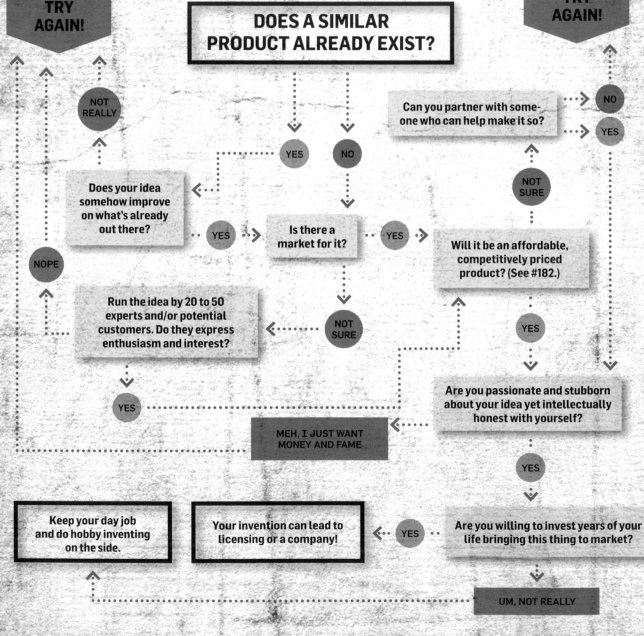

DOES A SIMILAR PRODUCT ALREADY EXIST?

TRY AGAIN!

TRY AGAIN!

NOT REALLY

YES

NO

NO

YES

Can you partner with someone who can help make it so?

Does your idea somehow improve on what's already out there?

YES

Is there a market for it?

YES

NOT SURE

Will it be an affordable, competitively priced product? (See #182.)

NOPE

Run the idea by 20 to 50 experts and/or potential customers. Do they express enthusiasm and interest?

NOT SURE

YES

YES

Are you passionate and stubborn about your idea yet intellectually honest with yourself?

MEH, I JUST WANT MONEY AND FAME

YES

Keep your day job and do hobby inventing on the side.

Your invention can lead to licensing or a company!

YES

Are you willing to invest years of your life bringing this thing to market?

UM, NOT REALLY

013 DISCOVER BRAINSTORMING

Though now in common use, *brainstorming* is actually a fairly new word. It was popularized by Madison Avenue ad man Alex Faickney Osborn in his 1953 book *Applied Imagination*. He was a successful executive and a prolific writer whose notion of brainstorming referred to a fairly specific method for collective problem solving that he first outlined in 1948's *Your Creative Power*. In an Osborn-style brainstorming session, a team of about a dozen people, including both experts and novices, would go through a step-by-step process of generating, sharing, and refining ideas in an organized group exercise.

Today there are dozens of variations on Osborn's process, including breakout brainstorming (participants divide into groups to generate ideas), Post-It brainstorming (participants write ideas on sticky notes for posting on a bulletin board), and telephone brainstorming (each participant writes down an idea and passes it to the next, who elaborates on it). And that's not counting the electronic approaches using special software and general-purpose tools such as email and social media.

014 BRAINSTORM WITH A GROUP

A classic brainstorming session can work wonders for your team. This method is based closely on Osborn's original book.

STEP 1 Gather a group of five to ten people in a comfortable room. Their backgrounds should be as diverse as is practical, especially with respect to their level of expertise in the subject: You want some wise old hands as well as some rank beginners. It's also a good idea to include at least two naturally gregarious folks who will keep the conversation going.

STEP 2 Present the group with a clear statement of a specific problem, rather than a general one. It should be the kind of problem you can talk through without having to do too much individual pen-and-paper work, though a chalkboard or whiteboard may be handy for explaining things to the group.

STEP 3 Present the four cardinal rules. It may be helpful to write them on the board.
- Go for quantity of ideas.
- Withhold judgment. Criticism happens tomorrow.
- Shoot wild. The crazier the idea, the better.
- Combine and improve upon earlier ideas.

STEP 4 Call for ideas, prompting discussion as needed to keep it lively. Gently remind people who stray from the rules. Keep a written list of all the ideas generated. When you're out of time, thank everyone for their participation.

STEP 5 Distribute the list to everyone in the group. Wait at least a day, then assemble a team of stakeholders to decide which ideas to carry forward. Now, and only now, is the time to put on your critic's cap and start sorting the wheat from the chaff. Before throwing out an idea, ask yourself if it can be modified or combined with another idea to fit the bill.

015 HONE YOUR DRAWING SKILLS TO DEVELOP YOUR IDEAS

Even now that desktop, laptop, and tablet computers are everywhere, many creatives still like to start out developing their ideas on paper. Nothing in the digital world really compares to the convenience and immediacy of a pad of blank sheets that you can scribble and tear through as your vision takes form. That doesn't mean you have to be a great artist (or even a good one) to use drawing effectively; mastering the few basic skills presented on these pages will get you where you need to be. As in any creative process, giving yourself permission to work quickly and without judgment at first is key. You don't have to be Leonardo da Vinci right off the bat (or ever), but it may help and inspire you to check out his and other famous inventors' notebooks.

Soon, you'll want to use a 3D-modeling program to make detailed digital models of your prototypes (see #021–024), but when you're still in the early-days brainstorming phase, relying on software will only slow you down. (Although many of the tips and tricks for sketching ideas on paper will serve you just as well when you start designing your 3D computer models.) At this stage, what you really want is an opportunity to refine your idea—to ask yourself questions about its shape, functionality, scale, and user experience, and to explore your initial answers to those questions via pen and paper. It's also a great way to consider the parts you may need—in fact, you can use your sketch to create a preliminary parts list, which will help you source materials for your prototype.

016 PICK THE BEST VIEW FOR YOUR SKETCH

The whole problem of drawing is representing the real 3D world on a flat 2D surface in a way that's easy to understand at a glance. By far the simplest method is to leave off the third dimension and find a way to present the subject flat, but sometimes you'll need to include perspective for context. Engineers, architects, mechanics, and others who make working drawings have come up with several standard approaches.

PLAN VIEW

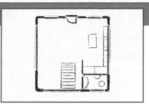

Strictly speaking, a *plan* is a 2D view of a location, structure, or object from above. Think of floor plans—everyone knows that they're looking down into a building that has walls and a roof that aren't shown. Finished plans are drawn to scale, meaning a certain distance on the drawing will always correspond to a certain distance in real life.

ELEVATION

An architect's term, an *elevation drawing* is a 2D view of an object (usually a building) from the front, side, or back, drawn as if seen from a distance so far away that the effects of perspective aren't noticed. This is also called an *orthographic projection*.

PRINCIPAL VIEWS

The exterior of any object can be shown with a set of six orthographic projections—one each of the front, back, left, right, top, and bottom. The front view goes in the center, the right view to the right, the left view to the left, and so on. The back view, if required, goes to the far left.

CROSS-SECTION

If the object is hollow or has interior details that can't be shown in one of the six principal views, add a sketch that shows what the object would look like if sliced open along a plane. Put this cross-section near the area that it offers a glimpse into, indicating where the subject has been "cut."

PERSPECTIVE

The *perspective view* accounts for the fact that, in the real world, an object appears larger close up than far away. These aren't often needed in initial sketches—unless you need to offer context by showing one, two, or three *vanishing points* (the specific spot where all parallel lines recede).

ISOMETRIC

Easier to make than perspective views, the *isometric projection* uses no vanishing points; instead, lines showing length, width, and depth are drawn at 120 degrees from each other. Another advantage is that faraway and near objects appear at the same size, while faraway details tend to become less legible in perspective drawings.

EXPLODED VIEW

The *exploded view* is a drawing, usually of all three dimensions, that shows how parts fit together in an assembly. Each part, or subassembly, is drawn separately and placed on the page to indicate how it fits with the others, often with lines connecting them.

017 DRAW A CONCEPT SKETCH OF YOUR INVENTION

Instead of trying to take an idea from blank slate to finished product on a single piece of paper, make lots of little drawings (called *thumbnail sketches*) that evolve toward the goal, working quickly at first and slowing down as you start to figure out what goes where. As you sketch, fight the urge to fix anything; just move over on your page or tear off a new sheet and redraw, keeping the stuff you like. The whole point is to get your ideas out as quickly as possible and not become bogged down with perfecting. To get started, you'll need a pen or pencil that feels good in your hand, along with both some cheap and nice paper. Find a clean, flat working surface with bright, even light, and of course have your recycle bin close by.

STEP 1 Even before you put pen to paper, you likely have an idea of your invention's size and shape. Is it a discreet handheld device designed for personal communication, or a large machine meant to push heavy stuff around? A rotating tray that helps dinner party guests help themselves to hors d'oeuvres, or a tall structure that aids in satellite communications? All these ideas—which are taken, just saying—have a shape that makes sense for their job. Start your sketch by trying out a few shapes that may make sense for yours.

STEP 2 Next up, consider what mechanical or electronic parts your invention needs to actually *do* its job. Does it need levers? Pulleys? Wheels or pumps? A motor or a speaker? Does it need to gather information from its environment via a microphone, thermometer, or some other input? You'll later do a ton of refining to work out these details, but for now—using your preferred shape as a base—draw a few cartoon versions of how your invention might achieve its purpose, including how the components attach and how they're sized in relationship to each other.

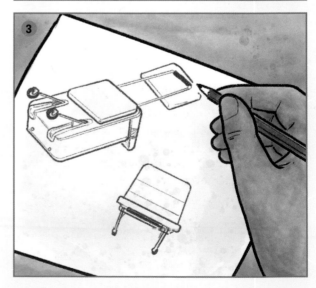

STEP 3 Consider showing multiple views. Chances are, some of your functions happen *inside* the invention, hidden from prying eyes and hands. It may help to work out these interior functions in an alternate view or a cross-section.

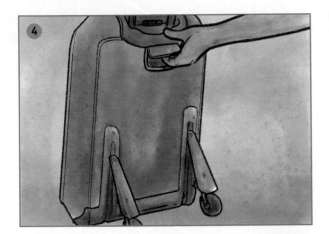

STEP 4 Once you have a rough idea for how your invention might do its thing, think about how the user will interface with it. Are there buttons? Dials? Is there a touchscreen? Where's the most convenient place for these interfaces to go on your chosen shape? How will they link up or communicate with the parts that do the invention's job? Draw a few drafts of possible interfaces and choose from those that you think are most intuitive.

STEP 5 Scale up. Your goal in evolving thumbnail sketches is to take a first pass at solving basic problems. Once you've got a sketch you like, switch to your nicer paper and lightly copy it out at a more legible size. A standard trick is to trace a 3x3 grid over the thumbnail, scale up the grid, and then fill in the boxes on the big grid one at a time. Blowing up your thumbnail will make obvious what important details are missing, so sketch them in now.

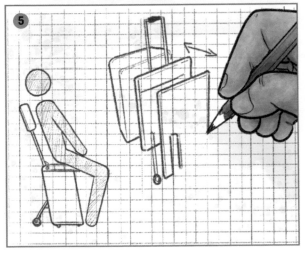

STEP 6 Write your name, the date, and the version number in one corner, add color sparingly if needed, and trace over your best lines in pen. Store in a binder after scanning the drawing at a reasonable resolution; 300 dots per inch (dpi) should be plenty. A digital file is useful as a backup in case your original gets damaged or lost.

018 TRY YOUR HAND AT CLASSIC DRAWING EXERCISES

Remember: We all drew as children. Even if you haven't drawn anything in years, dive in and try it. Being a good inventor likely won't make you a better artist, but being a good artist can make you a better inventor.

UPSIDE DOWN Find a line drawing that you like, turn it upside down, and copy it freehand. The point is to make your brain focus on the lines themselves, instead of what they represent.

STORYBOARD Show a simple action or story comic-strip style, as a series of panels that read from left to right. Imagine you're directing a movie and each panel is a separate camera shot.

ONE LINE Try to capture a subject, either live or in a picture, by drawing only a single line, never lifting your pen from the page. This is a hard one, so don't get discouraged by bad results. Just making the effort will do wonders for your skills.

019

MEET DR. NAKAMATS
INVENTOR OF THE PYON PYON BOOTS

Dr. NakaMats is, as they say, big in Japan. An 88-year-old inventor with more than 3,500 patents to his name, he's developed a cult following, regularly appearing on TV or online to demo one delightfully wacky creation after another. Take, for instance, his Pyon Pyon boots, a pair of running shoes mounted to springs that protect the body from impact during exercise. Or LoveJet, a mysterious treatment designed to inspire—and enable—lovers to do their part in replenishing the Japanese population. There's also a self-defense wig, a shaggy orange toupé equipped with a stealthy ninja throwing star. Users toss it (tresses and all) at unsuspecting foes, only to have it return—"just like a boomerang," NakaMats says.

While NakaMats's projects are undeniably playful, he's serious about being an inventor—and the persona that goes with it. He began his lifelong career in innovation when his mother enrolled him in science classes at the age of three. The jumpstart worked: Soon he was improving upon model airplane designs and patenting his own soy sauce pump—a kitchen gadget created for his mom.

From then, NakaMats never stopped making things. Today, browsing his Tokyo headquarters, you'll discover Ninger, a "car" that consists of a single roller skate propelled by a fuel cell, and a golf putter that makes a cheerful *PING!* when struck properly, among many other creations. NakaMats is also passionate about documenting his work—he's published a book that details his numerous patents and drawings. Even the shape of his office's door is a shoutout to one of his most proud—if most disputed—contributions: the floppy disk. (For the record, IBM owns the patent to the famous floppy, and they claim that they created it on their own. However,

continued on next page

they do acknowledge some licensing agreement with NakaMats, who is adamant that the tech giant approached him for the rights in 1979.)

As of this printing, NakaMats claims to hold the most patents in the world. "Thomas Edison only has 1,093," he boasts. "And I'm still living. And still inventing!" But how does one person continuously crank out so many ideas—each of them so unique? Perhaps unsurprisingly, NakaMats's creative process is as fascinating as its results. "I have a special room—a calm room in which all noises in the brain will be wiped out," he explains. "It's a golden room—all the walls are made of gold to shut out frequencies." He pauses, then delivers the punch line: "The toilet is also gold."

If a sit on this golden throne doesn't result in a sufficiently good idea, NakaMats has other tactics. One involves staying underwater "until 0.5 seconds before death," he says. "Very dangerous. But it is important to reduce oxygen in the brain to have extreme ideas." (To keep track of his thoughts when submerged, he invented a special waterproof pencil and memo pad.)

While it's not advisable to risk drowning to come up with a compelling concept, there's something to be said for developing a process that gets your creative juices going, no matter how ridiculous. It's equally important to devise a system for deciding if an idea has any real merit once the flash of inspiration passes. For NakaMats, an idea must have *suji* (a strong theory), *pika* (a flash of inspiration), and *iki* (practicality). According to NakaMats, "All inventors dream. But that is not practical. Inventors need to also think, 'How about this price? Is it suitable for this market, or is it possible to do it so it's good for the environment?'"

NakaMats's fandom is now worldwide, with several cities hosting annual celebrations in his honor, according to the man himself. In 2005, he received an Ig Nobel Award for photographing his meals for 34 years, all part of his study into which dietary choices spark the best ideas. (Salty foods seem to do the trick.) And while NakaMats surely courts celebrity, he describes a higher calling: "My spirit of invention is not about making money," he says. "It is about making people happy. The pump I invented as a child—that was made out of love for my mother."

The upshot? You may not need a gold-plated bathroom or an underwater idea chamber to come up with a game-changing invention. You may just need to pay a little attention to the people and problems in your life, and then craft a solution that sticks.

Q+A

Q: What's your favorite tool?

A: My hands. They are the best of tools. I can tell if I look at people's fingers if they're inventors—inventors have special fingers that can make anything.

Q: Are there any other inventors or makers who particularly inspire you?

A: My mother. She had a very creative mind. She was a great inventor. She invented me.

Q: Do you ever get stuck when you're working on an invention? What do you do to hit refresh?

A: I have a strict discipline. I never give up.

Q: Your inventions often have interesting names. How do they come to you?

A: When I was working on the floppy disk, I was listening to Beethoven #5. (That's the only music I can use to invent. Other music—such as Mozart—is not suitable.) Then a black butterfly came in through the window. It was flapping and waving in the air, and I decided to name the invention "floppy" because of its movement.

Q: What is your secret to success?

A: My theory is I never ask for money. If you ask people to invest, they have strong opinions. The inventor cannot move freely. That creates a not very good product. So I assume 100 percent risk, but I also have 100 percent freedom.

Q: Do you have any words of wisdom for aspiring inventors?

A: First, you should be honest. Second, you should study very hard. Third, do physical training. Inventing requires many long days that turn into long nights, and you must have a very strong body and spirit.

Dr. NakaMats's inventions range from the practical (a pillow that helps drivers stay awake at the wheel) to the far-fetched (a condom embedded with a magnet for increased "sensitivity"). While a healthy dose of skepticism is required—it is unlikely, for instance, that he invented the CD, DVD, fax machine, self-driving car, karaoke machine, and drone, as he asserts—the gadgets he *has* made are remarkable in their ingenuity.

Ever wish you could doze at your workplace with impunity? Don these glasses. The lenses are concealed by an image of a more alert set of peepers.

Always the first one to the party, Dr. NakaMats created the world's first "wrist phone" in 2003. He later updated his design to be smartphone-friendly.

Boost mental performance—or at least deter conversation—with Dr. NakaMats's Cerebrex armchair. Sound frequencies pulse throughout the chair in order to stimulate blood flow and increase activity in the brain. Nakamats himself enjoys a daily 29-minute nap under its hood, claiming that it improves memory, math skills, and eyesight while lowering blood pressure.

020 HEAR INVENTORS' THOUGHTS ON IDEATION

How do you invite eureka moments? Experts throughout the ages—and across disciplines—talk about where their ideas come from.

"The idea of Twitter started with me working in dispatch since I was 15 years old, where taxi cabs or fire trucks would broadcast where they were and what they were doing." *– Jack Dorsey, cofounder of Twitter and founder of Square*

"EVERYONE WHO'S EVER TAKEN A SHOWER HAS AN IDEA. IT'S THE PERSON WHO GETS OUT OF THE SHOWER, DRIES OFF, AND DOES SOMETHING ABOUT IT WHO MAKES A DIFFERENCE." *– Nolan Bushnell, inventor of Atari*

On embracing diversions: "Well, it was kind of an accident, because plastic is not what I meant to invent. . . . I was trying to make something really hard, but then I thought I should make something really soft instead, that could be molded into different shapes. That was how I came up with the first plastic. I called it Bakelite." – Leo Baekeland, father of plastics

"BE ALONE, THAT IS THE SECRET OF INVENTION; BE ALONE, THAT IS WHEN IDEAS ARE BORN." *– Nikola Tesla*

ON PAYING ATTENTION TO PROBLEMS BEYOND YOUR OWN BACKYARD: "WHEN I SEE SOMEONE STRUGGLING, I WANT TO DO EVERYTHING I CAN TO EASE THAT STRUGGLE. I'M INSPIRED BY EVERYONE AROUND ME, AND BY READING ABOUT WHAT THINGS ARE LIKE IN OTHER PARTS OF THE WORLD."
– Danielle Applestone, CEO of Other Machine Co.

"The air is full of ideas. They are knocking you in the head all the time. You only have to know what you want, then forget it, and go about your business. Suddenly, the idea will come through. It was there all the time." – *Henry Ford*

"I never really thought of myself as an inventor but a problem solver." – Ann Moore, inventor of the Snugli baby carrier

"ALL SORTS OF THINGS CAN HAPPEN WHEN YOU'RE OPEN TO NEW IDEAS AND PLAYING AROUND WITH THINGS." – *Stephanie Kwolek, inventor of Kevlar*

"ALL CREATIVE PEOPLE WANT TO DO THE UNEXPECTED."
– Hedy Lamarr, coinventor of a WWII radio-guidance system for torpedoes

"I'll see something that inspires me and I'll email it to everyone in the company. We all do this, so I'd say our innovation is a combination of people and process. We inspire each other and iterate. One of the most important truths we live by is that disciplines are dead, and the most interesting ideas are often at the intersections."
– Ayah Bdeir, founder of littleBits

021 DIVE INTO COMPUTER-AIDED DESIGN (CAD)

So you've got a rough sketch of your idea. Your next step is to translate it into a more rigorous, detailed digital 3D model that will help you create a prototype—for some, it helps to do this before you even *think* about picking up a screwdriver. No matter how good your spatial reasoning skills, you'll notice things in the digital model that you didn't see in your head beforehand, and the process of working up that model will solve lots of problems upfront. Plus, while the prototype is your physical goal, the most lasting and valuable output of the design process isn't a perfect one-off. Instead, it's the blueprint that you used to get there: the dimensions, tolerances, materials, and other technical know-how.

These days, the best way to develop and record this know-how is as a set of 3D *computer-aided design* (CAD) models. These files should represent every part that goes into the product, each accurate down to the last detail. When choosing a CAD program, remember that your goal is not to create video game art—you need *solid object modeling*. Popular commercial programs include SolidWorks, Solid Edge, and the Autodesk family of products. There are also easy-to-use, cheaper, or even free open-source programs: SketchUp is a good place to start if you're a WYSIWYG (what you see is what you get) person; OpenSCAD may be a better choice if you'd rather program your models than sculpt them on-screen.

022 GET THE FULL BENEFITS OF CAD

Fitting all your invention's required parts together so that it's functioning, good looking, and economical to manufacture is a lot like designing a 3D jigsaw puzzle. Luckily, solid object modeling software makes the process much easier, and offers other benefits too.

SAVE TIME AND MONEY Using a CAD program will make it easier to notice and correct assembly problems before they find their way into your physical prototypes.

SIMULATE REAL-WORLD USE Many CAD programs let you run simulations that can predict how parts and assemblies will respond under real-world conditions, and hopefully find and solve problems before you actually build anything. This is called *computer-aided engineering* (CAE).

IMAGINE THE FINISHED PRODUCT Your 3D CAD model can be used to produce so-called *renderings* of your product that look almost as realistic as photographs—before it even exists. You can also use it to export perspective line drawings, perhaps for use in a patent application or an instruction manual.

MAKE MOLDS AND OTHER TOOLING The parts required to make one unit of your product are divided into two categories: *off-the-shelf* and *custom* parts. The former you can buy and the latter you'll have to make or have made. The process of tooling up to manufacture your custom parts begins with the 3D CAD models. This is called *computer-aided manufacturing* (CAM).

023 LEARN HOW COMPUTERS THINK ABOUT SOLIDS

Making a computer understand a 3D object is complex. Basically, the machine needs to be able to quickly figure out if any point is inside or outside the model, so it needs to know where the object's boundaries are. Upshot: The computer wants to accurately represent the *surface* of the object, using one of three common methods.

MESH If you've ever seen the original *Tron* movie, you know what a 3D mesh model looks like: smooth surfaces that are approximated as a series of flat facets, like a gemstone. These facets are called *polygons* and consist of three basic features: *Vertices* are the points at the corners, *edges* are the lines connecting vertices, and *faces* are the areas enclosed by edges.

Mesh modeling is a powerful tool for computer graphics but has shortcomings for engineering. Imagine scaling up a small mesh object so it's ten times bigger. The surface may look smooth and work well at the small size, but when you enlarge it, the illusion vanishes: It's angular, faceted, and may not look or work like you want.

MATH Instead of defining a surface as a bunch of interconnected points in space, you can find a mathematical formula that exactly represents it. This trick eliminates the scaling problems of mesh modeling because a mathematical surface is always smooth. There are several tricks for making mathematical surfaces, with an alphabet soup of fancy names—like *non-uniform rational B-splines* (NURBs) and *constructive solid geometry* (CSG)—but they all boil down to this: Instead of telling the computer what the surface looks like, you teach it rules about how the surface is supposed to be, and let it use those rules to answer its own questions.

BOTH This method, called *subdivision surfaces*, is a mesh-math combo. Models are stored as meshes but with mathematical hints that help the computer add facets while keeping the same smoothness. There are various tactics for doing this, including the *Catmull–Clark algorithm*, the *Doo–Sabin method*, and the *Loop scheme*.

024 MAKE YOUR FIRST 3D CAD MODEL

If you're just starting out with CAD, try this classic 12-piece 3D puzzle before diving into modeling your invention. Called *solid pentominoes*, these shapes are easy to model, fun to play with, and good at getting your visual-spatial thinking muscles working. This project teaches the three basic *Boolean operations—union, subtraction,* and *intersection*—which you will use over and over again in 3D modeling to build up complex shapes and hollow out solid parts so they're lighter and cheaper to make. When you're done, you can play with the pieces in your CAD environment or 3D-print them for real-world use (see #069). All you need to start is a computer and solid modeling software. We used SketchUp.

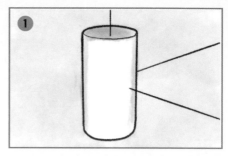

STEP 1 We'll start by modeling the negative shape that will be subtracted to hollow out the unit cube later on. The convention in 3D modeling is that X- and Y-axes are flat on the floor, while the Z-axis is up. Find the point where the three axes come together, called the *origin*, then use the circle tool to draw a ¾-inch- (19.05-mm-) diameter circle in the XY plane (i.e., on the floor) with its center at the origin. Now use the extrude tool to drag the circle 1 inch (25.4 mm) up along the Z-axis, sweeping out a cylinder. Click on the bottom face of the cylinder and extrude it down 1 inch (25.4 mm) along the Z-axis, this time to lengthen it.

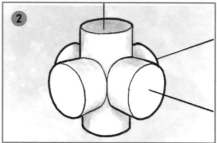

STEP 2 Select the whole cylinder, choose "group" so you can manipulate it as a single object, and copy and rotate it around the origin 90 degrees toward either the X- or Y-axis. Now you've got two cylinders that look like a plus sign. Copy and rotate the original cylinder again, this time in the perpendicular direction. Now you've got three cylinders that look like a toy jack. Select all three and make them into a group. This will be your mold group.

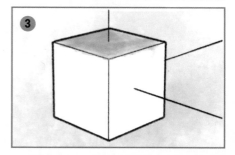

STEP 3 Next, we'll form the unit cube that we'll copy and paste to build up the pentominoes. Select the mold group and choose "hide." (It'll disappear, but you can bring it back anytime.) Grab the rectangle tool, click on the origin, and drag to make 1-inch- (25.4-mm-) square sides in the XY plane. Then extrude the square 1 inch (25.4 mm) up to make a cube. Now group all the parts of the cube together and use the move tool to transfer it ½ inch (12.7 mm) closer to the origin on all three axes, which should center it.

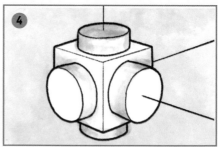

STEP 4 To prevent waste, we'll now remove material from the part's center. Start by unhiding the mold group. Your cube should seem to have a cylinder sticking out of each face. Make a copy of the mold group, put it in exactly the same place as the original, and hide the copy—we'll use it later.

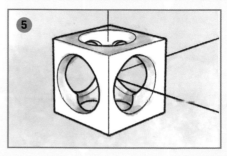

STEP 5 Select the visible mold group and run the Union operation to fuse all three cylinders together (much like redshirts caught in

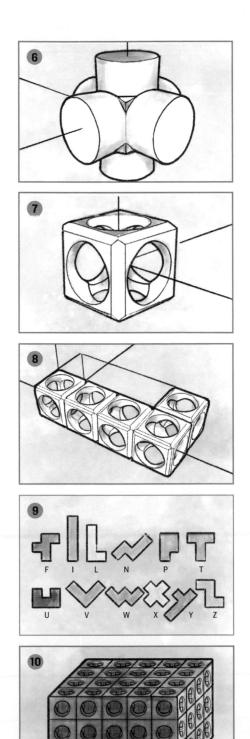

a transporter accident on *Star Trek*). Now select both cube and mold and run the Subtract operation, which takes a bite out of the cube in the shape of the mold. It does make a difference whether the cube is subtracted from the mold or the mold is subtracted from the cube; if you get something weird, undo the operation and try selecting the two shapes in the opposite order before running Subtract.

STEP 6 Next, we'll round the edges so our puzzle pieces don't have sharp bits. To do so, first unhide the mold group. It'll cover up your cube and look like a box with a cylinder sticking out of each face—much like step 4. Scale this mold up from the origin until your cube nearly fits inside the space where the cylinders converge, with just its corners sticking out.

STEP 7 Ungroup the three cylinders in the mold group, hide any two of them, select the remaining cylinder and the cube, and run the Intersect operation, which will cut off the bits of the cube sticking out. Voilà! Rounded edges! Repeat this step for the cylinders on the other two axes to round all the edges of the cube.

STEP 8 Now we've got a nice, hollow cube with rounded edges. This is a good time to save the file, naming it something like "unitcube." We just have to copy, move, and paste it a few times to make our first part. Start with the so-called L pentomino, which is four cubes in a line with one more hanging off the side at the end. Once you've got all five copies of the unit cube in place, select them all and run the Union operation again to fuse them into one part. Save the file as something like "L-pentomino."

STEP 9 To finish the set, load your unit cube model and repeat step 8 an additional 11 times to create the remaining pentominoes, saving each part as a separate file. Then create a new file called "pentominoes" and import each model file. Now you may play with them to your heart's content in the virtual world. You can also export each piece as an STL file and 3D-print it, or pay a service like Shapeways to print it.

STEP 10 Congratulations on your first 3D CAD model! We don't want to distract you too much from working on your invention, but might we humbly suggest that you tackle the classic problem illustrated at left? The challenge is to fit all 12 pieces exactly into the shape as shown.

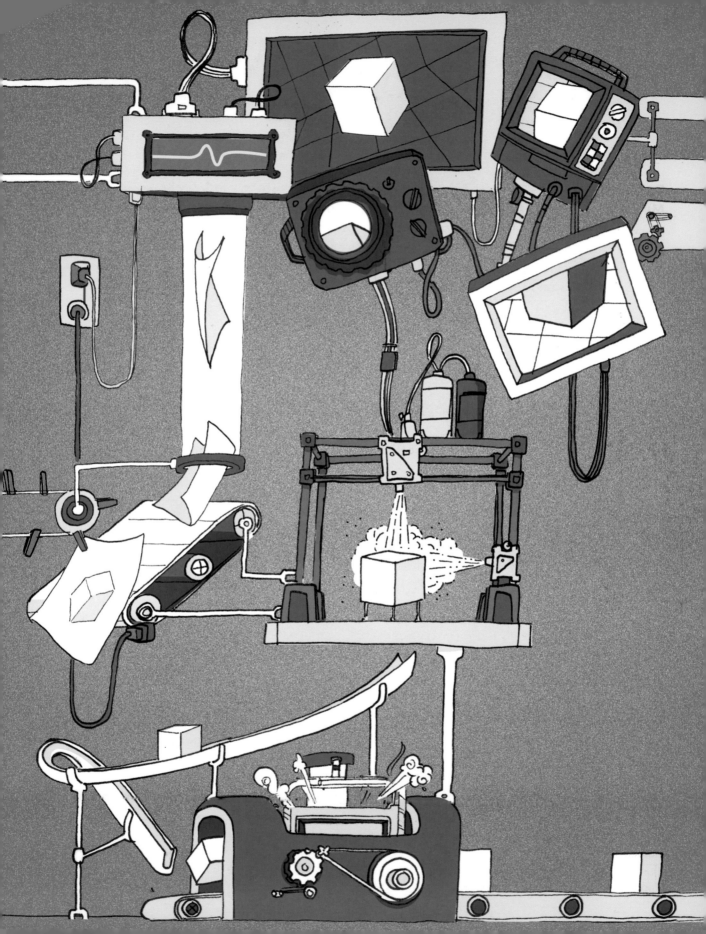

THE PROTOTYPING CYCLE

Believe it or not, every great product—from the light bulb, to the telephone, to the Apple Macintosh computer—started with a single prototype. And though there may be hundreds of versions between the slapped-together cardboard mockup and the shiny functioning model that rolls off the assembly line, every invention has to start somewhere. It may seem like a daunting proposition to finally begin putting solder to iron, or code to compiler, so in this section, we'll help break down the process into manageable tasks.

025 RIDE THE PROTOTYPING SPIRAL

In an ideal world, going from the first "eureka!" build of an invention to a mass-produced device for sale in brick-and-mortar stores would be an orderly, paint-by-numbers, straight-line process. But this ain't an ideal world and, truth be told, prototyping often feels like both a circle—in that you always seem to be starting over—and like a line, in that you're always moving forward. Every attempt brings the goal closer, though rarely as fast as you'd like. So let's call it a spiral, with each curve in the corkscrew marking a serious turn in your product's development. It's important to remember that these aren't setbacks—just bends in the road as you make uphill yet real progress.

026 WORK TOWARD MAJOR MILESTONES

Professional product developers learn to embrace the chaos of prototyping. Every product brings unique challenges, no two are ever exactly alike in their design cycle, and there's only so much you can do to control how things unfold. Nonetheless, there are major landmarks along the way to look forward to.

PROOF-OF-CONCEPT This build is a quick-and-dirty model that satisfies you, your team, and maybe your investors that it's technically possible to build a thing that does what you propose—and that it has a viable, real-world application.

WORKS-LIKE PROTOTYPE This version may be ugly as homemade soap, but it works in exactly the same way that the finished product is supposed to. This is the build that you'll start testing with users (see #057-065).

LOOKS-LIKE PROTOTYPE If you've got lots of capital and a big market, you may want to hire a designer to figure out what the thing should look like. She'll be trying to make the product as appealing to the eye as possible, and when she's done, there will probably be a full-size mockup available. (See #113–121 for more on this stage.) This prototype may not contain any moving or working parts, and it's designed according to marketing priorities, rather than engineering ones.

LOOKS-LIKE, WORKS-LIKE PROTOTYPE This version both looks and works exactly like the market-ready final product. Note, importantly, that it's still just a prototype and has not been through the process of design for manufacturing. More about that later; specifically, check out #129–142.

027 CHECK OUT FAMOUS PROTOTYPES

Even the inventions you use every day had humble beginnings. Just compare these prototypes to the famous products they eventually evolved into, and don't get all self-conscious yet. You can make it pretty later.

SUPER SOAKER At its inception, the world's favorite squirt gun was nothing more than few pieces of PVC pipe and a 2-L bottle held together with Plexiglas.

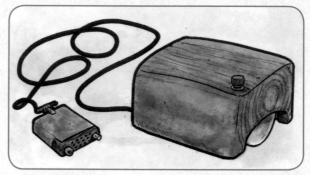

COMPUTER MOUSE Douglas Engelbart's design had a wood shell, two wheels that controlled a cursor, and a cord that resembled a tail—hence the name *mouse*.

TRANSISTOR Invented at Bell Labs in 1947, the first version of this game-changing electronic component was held together with screws and rough-cut plastic.

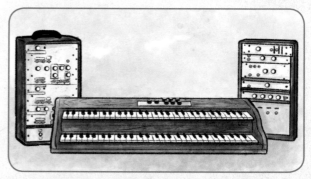

SYNTHESIZER Electronic music legend Bob Moog's prototypes often featured handmade circuits screwed to plywood and keyboards salvaged from organs.

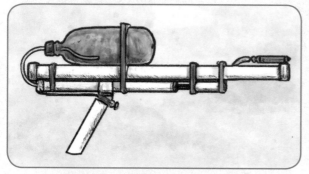

GOOGLE GLASS An early draft of the iconic eyewear consisted of a 3D-printed frame, hardware from a smartphone linked to an optical display, and a battery.

DYSON VACUUM In 1978, James Dyson took apart a vacuum and installed his own bagless filter, then built a quick, no-frills cardboard body for proof-of-concept.

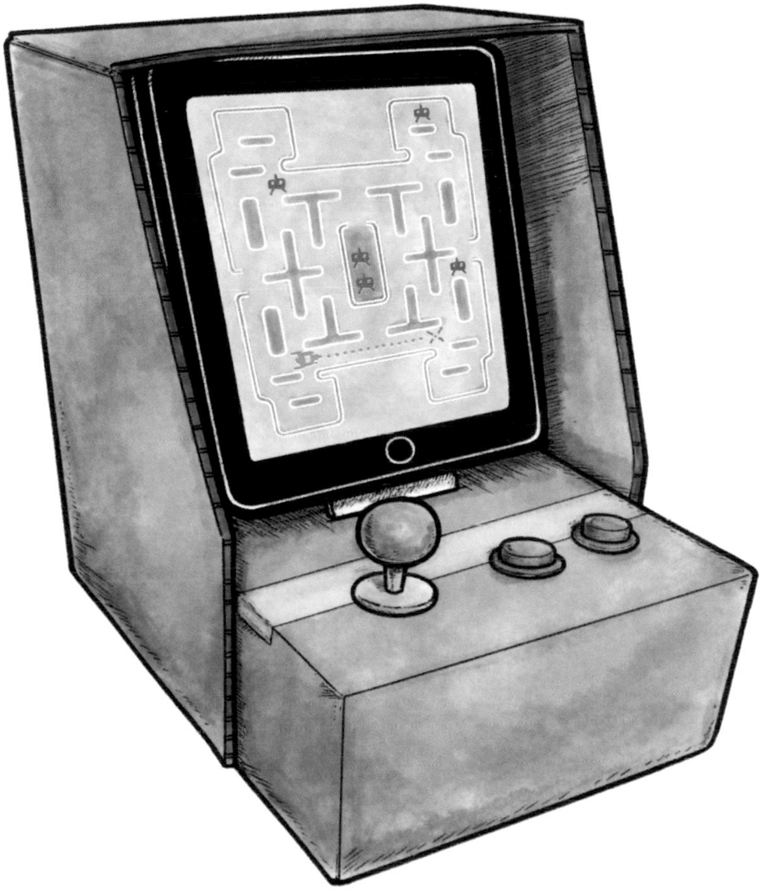

028 MAKE IT QUICK AND DIRTY

Right about now, your head may be exploding with possibilities: If this works, then so might this, and that, and these. It's easy to get overwhelmed—sometimes to the point where it's hard to start. If you get stuck, these guiding principles can help get you over the hump.

JUST BUILD IT ALREADY Don't get distracted by theory. Instead, cut to the chase and make a dead-simple, very minimalist working prototype—no matter how ugly—as fast as possible. Otherwise, you may get too hung up waiting on the tools, time, or materials that you think you need to make it perfect, which is impossible at this stage anyway. (This is also called procrastination. You know who you are.)

USE EVERYDAY MATERIALS Look around and you'll see that you're surrounded by stuff that, in a pinch, could be a useful material in your proof-of-concept build. Clay, wire, paper, PVC pipe, empty cardboard boxes (which you see above in our DIY arcade game)—these cheap and easy-to-work-with staples will set you free, allowing you to build almost as quickly as you can think without fretting about costs.

PLAY, DON'T PERFECT Use your prototype to see how well it works in a variety of situations before refining it. Learning with your hands will answer a lot of questions that you didn't even know to ask.

029 SAVE TIME WITH OFF-THE-SHELF BUILDING SYSTEMS

If you've ever played with a LEGO set, you already understand the basic idea of a building system: It's a bunch of premade, standardized pieces that can be arranged in infinite configurations. While kits like LEGO and Erector are fun to play with, makers often press them into service for the decidedly non-playful purpose of slapping together quick proof-of-concept prototypes for potential commercial hardware products.

There are grown-up versions of these building systems too, which are designed for scientists and engineers who need to put together tough working prototypes on tight timetables. LEGO has ventured more deliberately into this space with its Mindstorms kit series; their array of sensors and motors, combined with LEGO Technic bricks, have proven great in university and industrial robotics labs. And littleBits, while it began as an educational toy, is starting to come into its own as a tool for serious circuit prototyping (see #127). Finally there's 80/20 Inc., an industrial T-slot framing system that can help you build everything from enclosures to factory equipment to vehicle chassis.

030 TAP THE POWER OF YOUR PAPER PRINTER

Even the most complex of 3D shapes can be built up from flat slices. And while laser cutters and CNC routers are cheaper and more accessible than ever (see #040 and #149), another way to cut accurate parts from flat stock likely sits on the corner of your desk. I'm talking here about your paper printer, whether it's inkjet or laser. Even a low-end modern printer has a resolution of 1200 dots per inch (dpi) or better, which corresponds to a mechanical precision greater than 1/1000th of an inch (0.025 mm)—more than enough to mock up any device that's not a precision scientific instrument.

The point is that you could use software to design, say, a chair. Then, instead of tracking down a laser cutter, you can print out your design, stick it to your desired material, and cut it by hand. You achieve the same result—just without the fancy tool.

STEP 1 Design your part profile in a drawing program.

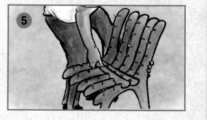

STEP 2 Print it life-size on one or more sheets of paper.

STEP 3 Affix these to the stock of your choice (cardboard, plywood, cardstock, etc.).

STEP 4 Carefully drill, cut, and file to the lines.

STEP 5 Assemble your pieces with any needed fixtures (screws, nails, and the like).

Sure, it can be a bit tedious, but it's reliable, handy, and accessible. A ream of full-sheet, printable, adhesive-backed mailing labels makes the process even easier: All you have to do is print, peel, stick, and cut.

031 SET YOUR PROTOTYPE ASIDE

Once your idea exists in the world as a prototype or model, put it down for a while. Take a deep breath and go on a little vacation from it. The point is to clear your mind and forget some of the trees so you can see the forest again. Depending on what kind of time pressure you're under, this could be as short as a coffee break or as long as a round-the-world trip. Just be sure to step back, as far as you can afford to, and get a little perspective. You'll be surprised by what you see.

032 COME BACK TO IT

This can be the hardest part. Once the initial euphoria of creation has dimmed, you have to switch modes and become the editor, the analyst, the critic. For a lot of people, creativity serves as an escape from boredom, and much of the satisfaction of the process is in the inspiration stage. Many of us would rather go on and have a new idea than go back and put in the perspiration necessary to refine an earlier one. But it's precisely this effort that makes the difference between mediocrity and excellence in the final product. Are you serious about being an inventor? If so, now's the time to prove it to yourself.

033 TRY AND TRY AGAIN

By now you should have an idea of what's working and what's not, and the challenge is to fix the latter without screwing up the former. You may have to go back to your empty room, limit your options, and redefine the problem. Now can be a good time to court serendipity and consider how your idea might have different applications for other users. If the Kalahari bushman in *The Gods Must Be Crazy* came upon your prototype in the middle of the desert, what would he think of it? What would he do with it? What would a child do with it? A convict? An architect?

The process of revision is, in truth, endless. You make one, you set it aside, you come back to it, then you make it better. Wash, rinse, repeat. The returns of this effort may diminish, or they may increase. At what point you stop is up to you, but keep in mind the words of da Vinci: "Art is never finished, only abandoned."

034 ASK THE BIG QUESTIONS

Once your prototype passes the first, most basic test—does it do what it's supposed to?—there are lots of other questions you should ask.

IS IT SAFE? If there's even a remote chance your invention could injure someone, fix it now. If the hazard is a side effect of its functionality (I'm looking at you, chainsaw), build in every possible safety feature and consider hazards from unintended or stupid misuses—and have a long, hard talk with your lawyer (see #099).

IS IT RELIABLE? Even if it works as intended 99 times out of 100, you've still got a problem. People won't recall the 99 times it worked, but they will remember—and kvetch about—the time it didn't.

IS IT EASY TO USE? In an ideal world, no product would require an instruction manual—the right way to use it would be obvious just by interacting with it. Ask people to use your product, observe how they get confused, and use that information to make improvements (see #057–065).

IS IT OBNOXIOUS? Even if it's as reliable and as a garden hose, a product that makes an annoying

sound or emits an unpleasant odor is going to put people off, generate bad buzz, and lose money.

IS IT GOOD LOOKING? Common sense here, known by every marketer and advertiser the world over: People like pretty stuff more than ugly stuff and are much more likely to buy the former (see #114).

IS IT SECURE? With the recent proliferation of Wi-Fi-enabled and connected devices, security has become a major sore spot. Testing is time-consuming and costly, the features often slow down a product's user experience, and most customers don't think much about security until long after the sale. But when breaches occur, they can create major publicity problems and cost tons of money. If your device is linked to the Internet, invest in protective measures upfront.

035 START LEARNING HOW TO MAKE STUFF

At the risk of stating the obvious, if you're inventing a physical thing—as opposed to something intangible like an app, a video game, or a brand—you'll need to know how to make real-world bits and pieces, especially during the prototyping stage when you're likely to be building stuff yourself. There are lots of companies out there who will gladly take money to build prototypes for you, and their services, though pricey, can be worthwhile if you're building something that requires a lot of expertise. But even if you manage to invent a physical product without ever picking up a tool yourself, you'll still need to understand the nuts and bolts of making stuff to make sound decisions about who's going to be doing the building for you, how much you're going to pay them, and how long it's going to take.

A physical invention is commonly referred to as a *device*. Devices are built from *assemblies*, which perform major functions—for example, a laptop manufacturer might speak of the display assembly, the keyboard assembly, or the hinge assembly. Assemblies, in turn, are made of subassemblies, which, I suppose, might themselves be made of sub-subassemblies, and these of sub-sub-subassemblies, and so on.

But at some point you get down to just parts, which aren't put together from anything else. And though there are literally thousands of tools, technologies, and techniques used to build stuff today, at the level of parts these processes can be divided broadly into three types—*additive, subtractive,* and *plastic.* (Note that plastic does not refer to the petroleum-based materials we often call by that name, but to the quality of plasticity—mashing or smushing a thing into a different shape.) Which type of process and which particular tools you use to make your parts will depend on many factors, and will certainly change as you move from to prototyping to production.

ADDITIVE Stock material is added to a workpiece to build up the finished shape.

SUBTRACTIVE Stock material is removed from a workpiece to carve out the finished shape.

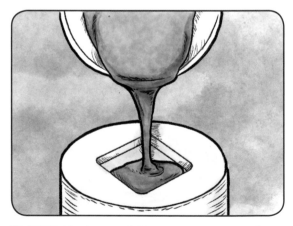

PLASTIC Stock material is moved around with pressure, heat, and/or chemical reaction to form the finished shape.

036 MATCH MATERIAL, METHOD, AND SCALE

There are many ways to make a part. Which you choose depends on production volume, materials, and facilities. Making a thousand on an assembly line demands different tools than making one on a kitchen table.

CRAFT SCALE

MATERIAL	ADDITIVE	SUBTRACTIVE	PLASTIC
CERAMIC	Coil pottery	Wheel trimming	Wheel throwing
WOOD	Model building	Carving	Basketry
METAL	Soldering	Engraving	Smithing
POLYMER	Polymer clay	Stenciling	Resin casting

SHOP SCALE

MATERIAL	ADDITIVE	SUBTRACTIVE	PLASTIC
CERAMIC	Figure modeling	Jiggering	Slip casting
WOOD	Cabinetry	Routing	Cold pressing
METAL	Welding	Machining	Spinning
POLYMER	3D printing	Laser cutting	Vacuum forming

FACTORY SCALE

MATERIAL	ADDITIVE	SUBTRACTIVE	PLASTIC
CERAMIC	Sintering	Grinding / Polishing	Press molding
WOOD	Truss fabrication	CNC routing	Steam bending
METAL	CNC welding	CNC machining	Die casting
POLYMER	Ultrasonic welding	Laser cutting	Injection molding

037 CRANK UP SOME MECHANISMS

When you build something with moving parts, you need driving energy. Mechanisms help by transforming one type of motion into another, or by carrying motion between parts of the machine.

GEARS Interlocking teeth prevent these wheels from slipping. Two or more can team up in a gear train to slow (and strengthen) or speed (and weaken) rotation.

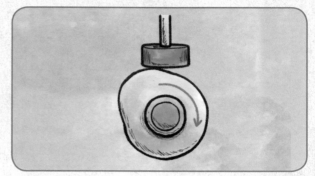

CAMS These rotating or sliding parts have an irregular tooth or bulge on an edge, which pushes against an accompanying piston or lever, called a *follower*.

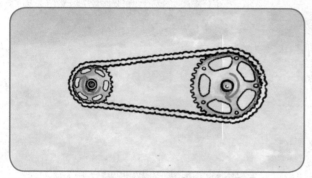

BELTS AND CHAINS These continuous, flexible loops are held under tension, transferring rotation from one sprocket, wheel, or pulley to another.

RATCHETS Wheels usually with teeth that, when combined with a small lever called a *pawl*, allow movement in one direction but not in the other.

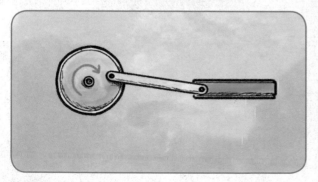

CRANKS Arms or wheels mounted on a shaft by off-center levers convert the shaft's rotational motion into the straight, back-and-forth motion of the lever.

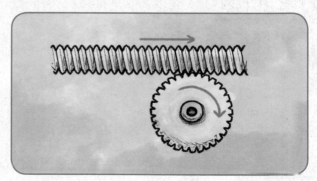

SCREWS These threaded rods push or pull along their lengths when turned. They're used as fasteners, linear actuators, and drive shafts (as in the "worm" shown).

038 KNOW WHAT YOU'RE GETTING INTO

While there are no absolute truths in predicting how challenging an invention will be to get off the ground, it helps to ask yourself two crucial questions at the get-go: Does it have electronic circuitry, and does it have moving parts?

THE DOORSTOP If you answered "no" to both questions, good news: You probably won't have to quit your day job. A product without electronics or moving parts is easy to prototype, so it's a good choice for a first-timer. Bonus: It's easier to sell a US$5 product made of one part for US$15 than a US$90 product made of ten parts for US$100, and it's the same ten bucks in your pocket either way.

THE BOARD-IN-A-BOX If you answered "yes" to question one only, your idea falls in the middle of the difficult-o-meter. Electronic devices with no moving parts have become surprisingly easy to develop. Today, they're largely assembled by integrated circuits (ICs) that take the headaches out of electronics troubleshooting. As long as your circuit is less complex than, say, that of a cell phone, it's possible to design it in software and have it work exactly as expected when built.

THE WIDGET If you answered "yes" to question two only, you've got yourself a mixed bag. Two or three moving parts may be pretty easy, between three and a dozen not too daunting, and above a dozen challenging. Prototyping complex machines is a necessarily long, involved, and expensive process. It may require starting a small company.

THE ROBOT If you said "yes" to both questions, go ahead and put in your two weeks' notice and hire a team. Best case scenario, the moving parts don't need to interface with the electronics, making it about as complicated as your board-in-a-box. But if your device has *actuators*—moving parts controlled by electronics—things get trickier. If those actuators need to move with great precision, they get trickier still. And if the parts in question are large, heavy, or fast-moving, it's dang tough.

039 ASSEMBLE AN INVENTOR'S TOOL KIT

Sure, your first prototypes can be held together with shoelaces and Bazooka bubble gum, but at some point do yourself a favor and invest in good tools.

SECURITY SCREWDRIVER SET The important thing here is to get all the right shapes—flat, Phillips, hex, Torx, Tri-Wing, Pentalobe, and more—and a handle with which to turn them. Save money and space with interchangeable bits and a matching driver.

POP RIVETER You can join two sheets of plastic or metal with a *pop riveter*. Insert the rivet into a premade hole in both materials, then use the gun to pull up on the rivet's mandrel, causing the wider shank on the bottom to expand and hold the two sheets together. It gets its name from the fun *POP!* it makes when the mandrel breaks off at a premade breaking point.

DRILL PRESS Bore perfect holes through metal, wood, or other materials with this shop classic, which comes in both power- and hand-feed versions. An array of bits lets you work with different materials.

SOLDERING IRON AND SOLDER If you're working on an electronic device, breadboard will get you through the first rounds. Eventually, you'll want to attach the components and wiring more securely, and that's where a *soldering iron* comes in handy. You heat its tip to melt and apply *solder*—a metal alloy that bonds other metal workpieces (see #147).

WIRE STRIPPERS This hand tool helps you peel back the rubber on wires, allowing access to the bare metal for easy soldering or twisting.

MULTIMETER When working with electronics, use this tool to troubleshoot circuits and measure voltage or current.

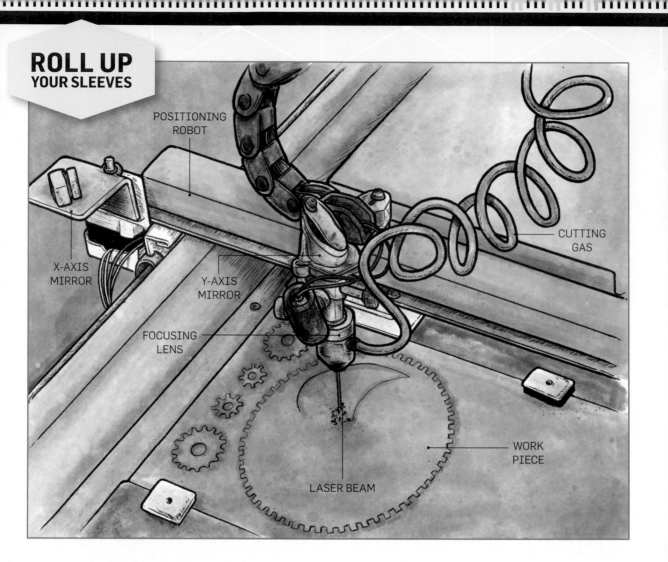

POSITIONING
ROBOT

CUTTING
GAS

X-AXIS
MIRROR

Y-AXIS
MIRROR

FOCUSING
LENS

WORK
PIECE

LASER BEAM

040 MEET THE LASER CUTTER

Pew pew pew! For those of you who've been living under a rock since 1960, a laser is not just a science-fictional death-ray weapon but a real-world device that projects a beam of coherent, tightly focused light in a very straight line over a long distance. It's an acronym for "light amplification by stimulated emission of radiation," but the word *laser* has long since come into everyday use as a common noun.

Lasers make it possible to project energy very precisely and accurately. At low powers in the optical spectrum, a laser is handy for occasionally pointing things out at a distance (but arguably makes a more amusing cat toy); at high powers in the infrared spectrum, it can act as a very fast, efficient cutting tool for materials ranging from paper to hard steel.

Laser cutting is a subtractive process, commonly used on flat stock like plywood and sheet plastic. In most laser-cutting machines, you'll find the laser housed in a safety cabinet, along with the cutting bed and the material to be cut. The laser itself is usually stationary, while mirrors direct its beam by moving along X- and Y-axes using precision mechanics and stepper motors. The cutting head often includes a lens mechanism that can change the beam's focal length to any point within the cutting depth. A stream of air blasts the head, clearing gases and bits of molten material away from the cutting area.

Now that you're acquainted, let's explore how you can harness this supertool to aid and abet in the process of prototyping parts for your invention.

> "The 3D printer gets 'em in the door. The laser cutter keeps 'em coming back."
>
> — *hackerspace maxim*

041 FIND ONE IN THE WILD

Today's laser cutters are pretty dreamy, and their ever-decreasing price tags and ease of use make them an increasingly tempting purchase for gearheads and novices alike. But don't clear out the guest room just yet: There are lots of ways to get access to a quality laser cutter, and—depending on your project—renting or using a service may save you in the long run.

BUY YOUR OWN As with 3D printing (see #066–070), a lot of formerly industrial-type laser-cutting technology has started trickling down into the consumer or so-called prosumer market range. Depending on your material, the size of the stuff you'll be cutting, and your intended volume of parts, it may be worthwhile to spring for your own laser cutter. As of late 2016, small desktop models capable of cutting ¼-inch- (6.35-mm-) thick acrylic up to a 10-inch (25-cm) square could be purchased new for just under US$2,000.

BORROW TIME If that doesn't cut it for you (sorry), there may be a hackerspace, makerspace, community college, or other tool-sharing collective nearby that will let you make use of a laser cutter, after a bit of training, in exchange for a recurring or one-time fee. You may also have a friend or an acquaintance with access to a laser cutter who may be willing to cut a small number of parts for you as a favor. So tapping your social network may be worthwhile, if your needs are modest.

CONTRACT IT OUT Finally, there are a large number of custom laser-cutting services in the world, including many you can access online: Design your part in software, upload it through the web, pay the piper, then wait a few days for it to show up in the mail. Ponoko.com is a good place to start if you need flat parts in a softer material like plywood, plastic, leather, or cloth. Need to laser-cut a round surface (such as a pipe) or work on something made out of a hard material like metal, glass, or ceramic? That equipment does exist, though it is so expensive that you'll almost certainly be paying someone else to run it for you, rather than owning or operating it yourself.

042 MAKE YOUR FIRST LASER-CUT PARTS

Laser-cutting machines almost always support two basic operations: *cutting* (obviously), in which the beam slices all the way through the material, and *engraving* (or *etching*), in which the beam just marks the surface. Try your hand at this starter project, which will help you understand these functions from software design to tangible prototype.

STEP 1 Make your part drawings with a program that edits vector art. There are several software options here: AutoCAD or LibreCAD, for engineer types who are comfortable using a computer-aided design program (see #021–024); Adobe Illustrator and CorelDraw, for graphic designers; and Inkscape, for folks who don't want to spend any money or want to support free, open-source software. It doesn't really matter what program you use, so long as it A) allows you to lay out your parts at final real-world size, and B) can save, export, or print to Adobe PDF format.

STEP 2 Preprocess your parts. To engrave an image, you'll need a program that edits raster art, like digital photos. Use it to convert your image to black and white or grayscale to make sure it looks okay, then adjust it to final size and position it in your drawing file. If you want to engrave vector art, you need to tell the laser cutter which lines to cut and which lines to engrave. Most machines will cut any line that has a stroke width of 1/1000th of an inch (0.025 mm) and will engrave everything else, but check before proceeding. Likewise, the color you use for lines or images may make a difference; default to black but read the instructions before hitting print. Once your design is ready, save, export, or print it to PDF.

STEP 3 Cut the file. If you're operating your own machine, it's worth saying that you should carefully follow the instructions that came with it to avoid damaging the machine. If you're borrowing time on a shared unit, you probably underwent training, to which you should of course adhere. If you're contracting the work out, this step amounts to uploading a file, emailing somebody, or handing over a USB drive. When it comes time to actually start the cut, it's usually as simple as positioning the material, closing the cabinet, and sending the file to the machine as though it were a printer.

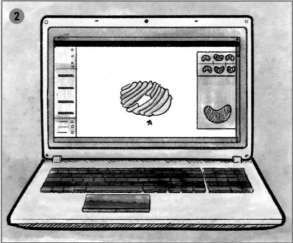

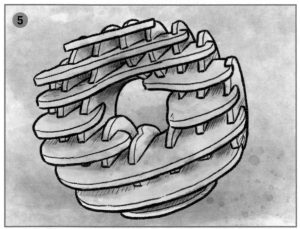

STEP 4 Postprocess your parts. Some flat stocks come attached to a thin carrier film on one or both sides. When the cutting is complete, you may find that you have to punch out the parts to separate them from the stock—sort of like die-cut cardboard game pieces. Depending on the material, there may also be a smell, which will dissipate with time. If you're working with acrylic, your parts should come out with nice, smooth, heat-polished edges. If you're working with a biologically derived material like cardboard, plywood, or leather, the edges of the parts will probably be charred. Some people find this annoying, but many see it as an aesthetically interesting effect—like artistic woodburning.

STEP 5 Slot, glue, or screw together (or otherwise assemble) your laser-cut parts to bring your prototype into three dimensions. There are lots of clever ways to build 3D solids from flat cutouts. Besides the grid of sections approach shown here, remember that you can also stack the sections up like layers in a topographic model. There's also the so-called *graphic profile technique*, in which complex parts are carefully designed to assemble into the finished form with a minimum amount of waste. This approach is the most efficient materials-wise, but it also requires the most time, attention, and effort from the designer.

043 DON'T SHOOT YOUR EYE OUT

Actually, that's dang unlikely with a modern commercial laser cutter. The cutting area should be completely enclosed, and the door should have a safety interlock switch that prevents the laser from operating when it's open. Nonetheless, be aware that these lasers can burn your skin and permanently damage your eyes, so respect them accordingly. Your machine should also have a big red emergency stop button that you can slap if something starts to go wrong.

The most crucial rule is to never laser-cut vinyl or any other plastic that contains chlorine; it'll release toxic chlorine gas, which can ruin the machine even if the area is safely ventilated. Metals, glass, and ceramics are also usually unsuitable for nonindustrial machines.

044

MEET STEVEN SASSON
INVENTOR OF THE DIGITAL CAMERA

Steven Sasson was a 23-year-old engineer at Kodak when his bosses asked him to determine if a new *charge-coupled device* (CCD)—a sensor that converts light patterns into electrical signals—had any practical value for the company. As a kid, Sasson used to build electrical gadgets in his basement, but he was no professional inventor. And he certainly didn't imagine that his tinkerings would lead to one of the most celebrated inventions of the 20th century: the digital camera.

Sasson wanted to capture images from the CCD, but that was impossible because it could only hold the light pulses it measured for a few milliseconds. As a fix, he decided to create a machine that would translate the pulses into numbers—digitizing them—and also store and display them. "I didn't want to digitize the world," he says. "I only decided to take a digital approach because, quite frankly, it eliminated some technical problems for me."

A trip to the local library, however, revealed that he would be on his own for creating such a device. "I couldn't find any ideas or approaches to doing this type of thing, so I just started trying to figure out how to accomplish this myself," he says. During the next year, he designed the camera and display, piece by piece, with cobbled-together parts, including a playback digital cassette tape system that hooked up to a television; an analog-digital converter; a lens repurposed from an old movie camera; and dozens of circuits. "I had to be creative because I didn't have many resources," Sasson says. "I begged, borrowed, and stole parts from Kodak."

In December 1975, it was finally time to see if he had anything to show for his efforts. He and Jim Schueckler, a technician who had helped with the project, started searching for a first subject to photograph. "We found a young lady named

continued on next page

Joy Marshall who happened to be working near our lab, and I asked if we could take her photo with this odd-looking contraption," Sasson recalls. "She knew us as the weird guys down the hall."

Sasson took a head-and-shoulders shot, positioning Marshall against a light-colored wall. While the actual exposure took just $1/20$th of a second, storing the image on the camera's magnetic tape took a whopping 23 seconds. When it was done, Sasson took the tape out and put it into the table playback unit, which—after 30 seconds or so—displayed the 100-by-100-pixel black-and-white image. The outline of Marshall's hair and the white wall were both visible, but her face was a jumble of static. Still, the image was beautiful in Sasson and Schueckler's eyes. "We knew a thousand reasons why we wouldn't see anything at all," Sasson says. "So we were quite happy to see something recognizable, since that meant that most of the circuits and software were working." Marshall, however, was less impressed. As Sasson recalls: "She took one look, said 'It needs work,' and walked out."

Luckily, however, the fix was relatively straightforward: Sasson had mixed up the order of bits, accidentally reversing the 16 levels of gray. An hour later, Marshall's face emerged from the static. Kodak's higher-ups, however, did not immediately embrace the possibilities of Sasson's invention. They protested that no one would want to view photos on a television screen and that the electronic image quality would never equal that of film. Over the years, Sasson saw many of those purported limitations fall away, to the point, he says, "that I almost smile about all the reasons I was told why that camera in your pocket or purse would not exist."

Q: What's your favorite tool?

A: My favorite tool is my pair of needle-nose pliers, which I've had since I was a kid.

Q: What's your earliest memory of tinkering?

A: I remember building a little relaxation oscillator circuit when I was just a kid. It used the nonlinear properties of the neon bulb to make a circuit that would light each of ten neon bulbs in a random order. I thought it looked like those early computers in the science-fiction movies of the early 1960s.

Q: What's it like to see your invention in the wild? Are you ever surprised by its uses?

A: I am surprised at how far the quality of images created by digital cameras has come. Also, the sizes of the cameras are much smaller than I ever thought they would get—at least in my lifetime. I never guessed the selfie would be a customer want.

Q: If you could go back through your invention process, what would you do differently?

A: I would try to communicate my vision more effectively to those funding my work. In the early stages, you simply don't have enough data to convince people that your idea has merit. I realize now that you have to appeal to people's curiosity and instil in them confidence that the problems in the way can be solved, even though the solutions may not immediately be obvious.

Q: Do you have any words of wisdom for aspiring inventors?

A: Always consider the audience who you are trying to convince of your invention's merit. Inventors live with their idea much longer than the audience they are trying to influence. Be patient with the questions and challenges you get. They're a sure sign that you are being heard.

Believe it or not, the sleek digital camera you now tote around was once composed of four distinct pieces: the camera (hacked together with a lens from a Super 8 movie camera, a cassette recorder, a CCD imaging array, and about a dozen different circuits); a cassette playback unit; a microcomputer that read the data from the cassette; and a monitor that displayed the data as an image on the screen.

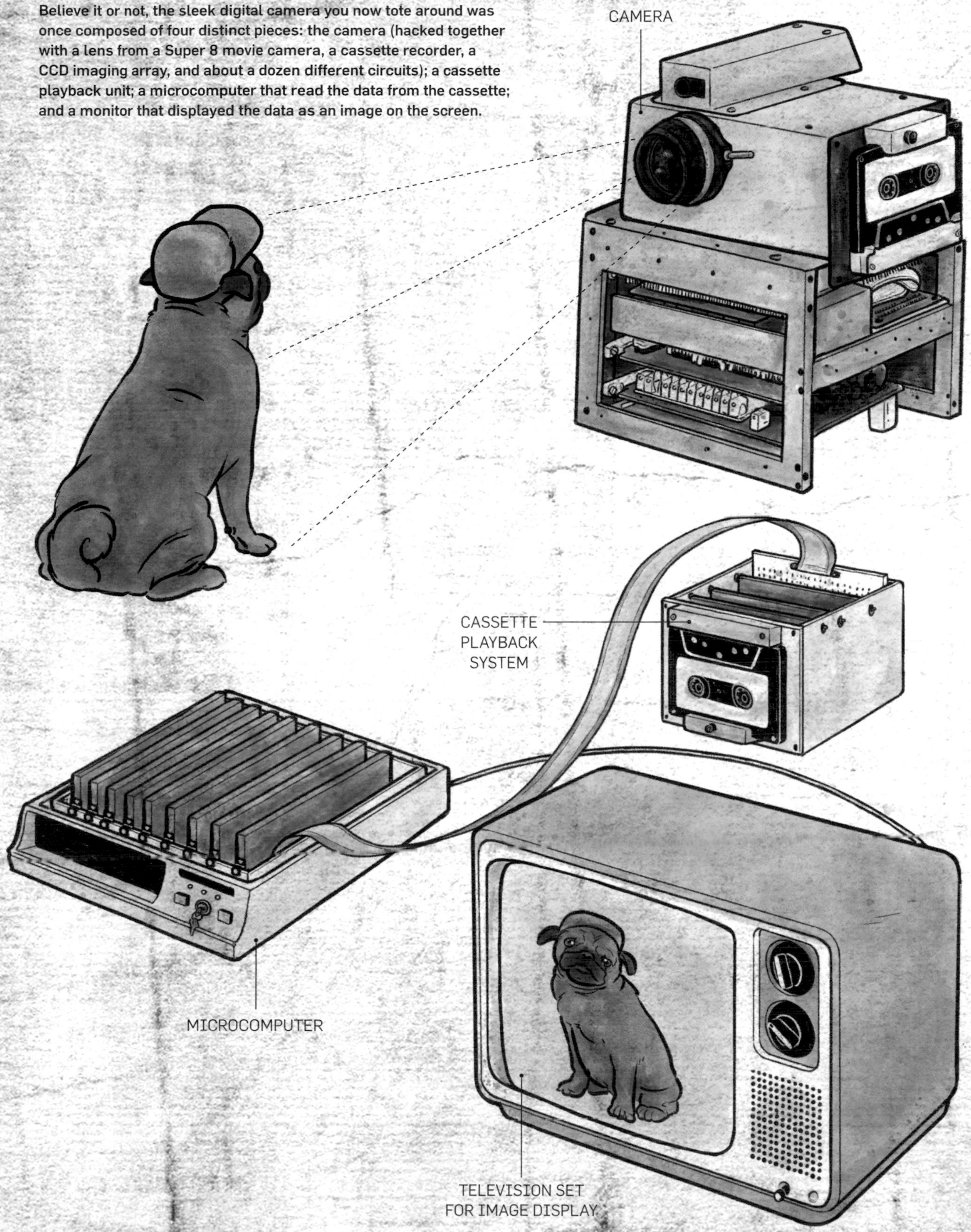

CAMERA

CASSETTE
PLAYBACK
SYSTEM

MICROCOMPUTER

TELEVISION SET
FOR IMAGE DISPLAY

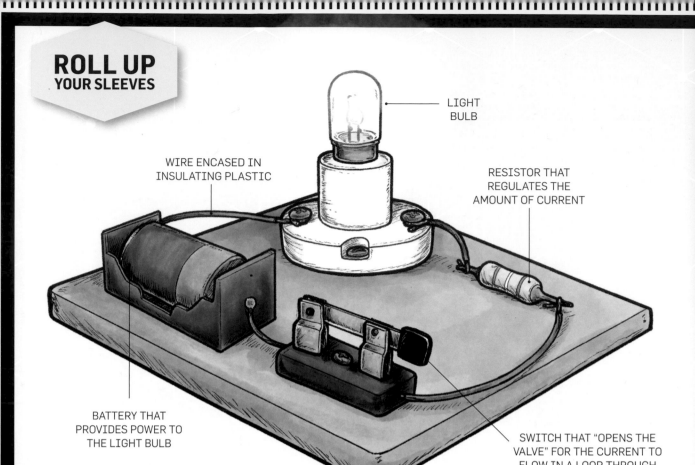

LIGHT
BULB

WIRE ENCASED IN
INSULATING PLASTIC

RESISTOR THAT
REGULATES THE
AMOUNT OF CURRENT

BATTERY THAT
PROVIDES POWER TO
THE LIGHT BULB

SWITCH THAT "OPENS THE
VALVE" FOR THE CURRENT TO
FLOW IN A LOOP THROUGH
THE COMPONENTS

045 GET THE BIG-PICTURE VIEW OF ELECTRONICS

Believe it or not, just growing up in the modern world has already taught you a lot about electricity and how it behaves. You almost certainly know that it can be lethal—we cover wall outlets to keep baby's fingers away and sing songs to older children about the dangers of downed power lines. And you've likely petted a cat on a dry day and felt a spark jump from your fingertip to the next object you touched.

But what's electricity all about? Hop in the wayback machine for a moment and remember sixth-grade science. The solid, liquid, and gaseous stuff all around us is made of unimaginably small bits called *atoms*, which in turn are made of even smaller bits called *protons, neutrons*, and *electrons*. Electricity involves only protons and electrons, which carry equal and opposite units of charge—protons positive, electrons negative. All the protons are jammed up in a tight little ball at the atom's center, with the electrons

orbiting around in a cloud. When an electron jumps on or off the cloud, the atom suddenly has more electrons than protons, or vice versa, in which case it becomes either negatively or positively charged.

These charges are what electricity is all about: They're the reason circuits light up, make sounds, move around, or sense their environment. Certain materials, like metals, share electrons among their atoms so easily that charge basically flows through them like water in a pipe. We call these materials *conductors*, and often shape them into wires to move charge from one place to another. Such a flow of charge in a conductor is called an *electrical current*.

Today many of the hottest new products are electronic. Learning to make basic circuits will get you started prototyping these kinds of devices and help you understand the science behind them. The crucial thing, especially at first, is to develop intuition.

046 IMAGINE WATER FLOWING

The idea that electrical current flowing in a wire is like water flowing in a pipe is a powerful tool for developing intuition about electronics. A flow of water, whether it's a river or a jet from a Super Soaker, has two fundamental qualities: There's the amount of water that flows through in a given period, and there's the pressure at which it flows. A lazy river at a water park moves a lot of water, but the pressure is low. A Super Soaker, however, doesn't move much water, but the pressure is high, which is why it shoots so far.

These two concepts (volume of water and pressure of flow) correspond nicely to electrical *current*—the flow of charge through matter—and *voltage*—the difference in potential energy between two points. Remember that this is just an analogy; nothing physically moves through a wire carrying current. But thinking about electricity this way will help you envision how various components will affect current as it moves through a circuit—and how you might use them to build working prototypes.

047 LEARN ELECTRICAL TERMS

Electrical know-how requires some jargon, most of which can be expressed by mathematical equations. But you don't have to think about the math to learn a lot. A quick vocab primer will help these concepts stick.

CURRENT

As we've mentioned, a current is a flow of charge through matter. A current that keeps reversing direction so the charge wiggles back and forth, instead of flowing straight through, is called an *alternating current* (AC). The wall plugs in your house deliver alternating current; most large appliances are designed to use it. A current that doesn't alternate is called a *direct current* (DC); most small appliances are designed to use it. Current is measured in units called *amps* with a tool called an ammeter (or multimeter). (See #039.)

COMPONENT

A component is a device with two or more points of electrical contact (called *terminals*) that affects electrical currents. Batteries and light bulbs are examples of components.

RESISTANCE

The tendency of a material to resist the flow of current, i.e., to be a bad conductor. A component designed to do this is called a *resistor*. Resistance has units of ohms (Ω), milliohms (mΩ), and so forth, and can be measured with an ohmmeter (or multimeter).

CHARGE

A property objects can acquire that causes them to be pushed or pulled away from other charged objects. Charges can be positive or negative. Positive and negative charges attract each other; two charges of the same type push each other away.

VOLTAGE

The difference in electrical "pressure" between two points. If you touch the two probes of a voltmeter to opposite ends of a new AA battery, you'll measure 1.5 volts between them. It has units of volts (V), millivolts (mV), and so forth, and can be measured with a voltmeter (or multimeter).

CONDUCTOR

A material that allows current to flow through it. Most metals are good conductors, which is why we make electrical wires from them. Most plastics are bad conductors, which is why we wrap plastic tubing around wires to prevent accidental electrical contact with the metal inside.

CIRCUIT

An arrangement of components connected in one or more closed loops so that current can continuously flow through the loop. Circuits can be designed to do useful things—for instance, in a flashlight where the circuit converts energy stored in the battery into light.

048 GET COMFORTABLE WITH COMPONENTS

Circuits are made by electrically interconnecting small devices called *components*. Components, even of the same type, come in an amazing variety of shapes and sizes, called *packages*. The important thing is not to memorize their appearances but to understand the fundamental categories, how they work, and their uses.

In general, there are two types: *passive* and *active*. Passive components aren't capable of power gain and can't supply energy into a circuit, while active components can add energy to a circuit and be used to control one flow of current with another. Within these two groups, there are basic component types that regularly show up in circuits, big and small, which we've shown below.

Here, the water analogy, which describes electrical behavior in terms of plumbing, can be extremely helpful (see #046). Get to know these parts and the roles they play in circuitry, and soon you'll be trying your hand at building your own.

PASSIVE COMPONENTS

ITEM	DESCRIPTION	PLUMBING ANALOGY
WIRE 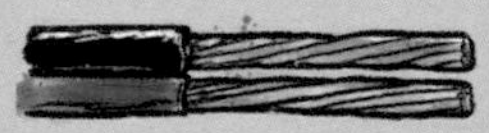	A length of conductor, almost always metal, that carries current.	Think of a wire as a pipe and the charge it carries as the water inside the pipe.
SWITCH 	A mechanical device that can interrupt the current in a wire. You use dozens of them every day: keys on your keyboard, the ignition in your car, and the light switch by your door.	The switch acts as a valve, blocking the pipe when you want to stop the flow.
RESISTOR 	A two-terminal device that limits the current flowing through it.	You can think of a resistor as a narrow spot in the pipe.
CAPACITOR 	A two-terminal device that can temporarily store a relatively small amount of electrical energy.	A capacitor acts like an elastic membrane stretched across the inside of the pipe. It will swell up to a point and spring back, given the chance, but it can "burst" if you push it too far.

ACTIVE COMPONENTS

ITEM	DESCRIPTION	PLUMBING ANALOGY
BATTERY 	A two-terminal device that can store a relatively large amount of electrical energy in the form of chemical energy for a long time.	A battery acts like a pump that takes water in at the bottom and then pushes it back out at the top.
DIODE 	A two-terminal device that allows current to flow in one direction only. There are many types, including *light-emitting diodes* (LEDs), which visibly glow in operation.	A diode functions as a one-way or check valve that prevents backflow along a pipe.
TRANSISTOR 	A three-terminal device that allows one current to control another, which may be much larger. Though there are many types, a good starting definition is "voltage-controlled switch."	Think of a transistor as a valve that opens or closes depending on how much water pressure is applied to a side arm.
INTEGRATED CIRCUIT 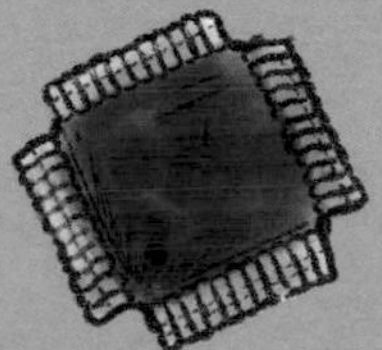	A prepackaged electrical circuit manufactured on a tiny chip of silicon. A modern IC can have dozens or even hundreds of terminals.	The IC is like a black box full of pipes and valves with input and output connections on the outside. You don't have to understand how it works inside, just how it will respond to input.

049 BUILD YOUR FIRST CIRCUIT

In the past 100 years, we've gone from stringing circuits together by hand on wooden boards to manufacturing billions of them every year using robots. The first transistor was the size of an egg; now we can fit billions of them on a chip the size of a quarter. Design, layout, and manufacturing of circuit boards is done using a sophisticated chain of software tools and industrial machines that require little physical human labor between the time the engineer provides the design file and finished boards roll off the assembly line. Even so, the best way to start learning about electronics is still by rolling up your sleeves and playing with a handful of components. Let's get to it.

For this simple circuit, you'll need only two tools: a solderless breadboard and sets of wire-cutting and stripping pliers. You'll also need the following components: a 9-volt battery and battery clip, jumper wires, a switch, a blue LED, a 220Ω resistor, and a 1000-uF electrolytic capacitor.

STEP 1 Start by adding the resistor, LED, and switch to the breadboard as shown. You may have to bend some of the component leads with your pliers to get them in the breadboard holes. Make sure you've connected the LED's long leg (aka positive lead or cathode) to the resistor.

STEP 2 Attach the battery to the clip and connect the leads to the buses. Add short jumper wires to connect the resistor and switch to the power and ground buses, respectively.

STEP 3 Flip the switch, and the LED should turn on. Now try it with the LED leads reversed and observe that it doesn't light up; that's because it's a diode and current can only flow through in one direction.

STEP 4 Add the capacitor to your circuit. Make sure the lead marked with the + symbol is the one connected to the resistor. When you flip the switch, the LED should turn on as before.

STEP 5 Now remove the battery. Instead of going out immediately, the LED should slowly dim away. When the battery is connected, the capacitor charges up; as soon as it is disconnected, the stored charge drains away through the LED. Pretty cool!

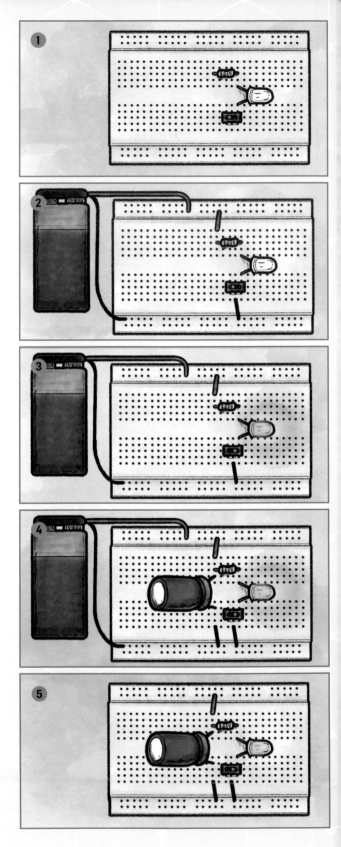

050 REV UP THE ACTUATORS

If you're building a robot or other widget that moves, you're going to need one or more actuators to turn electrical power into motion. Actuators are types of *transducers*, which are components that turn one kind of energy into another. Many actuators can be used backward in terms of energy—for instance, instead of applying current to turn a motor shaft, you can turn the motor shaft to generate current.

MOTOR

Motors convert direct or alternating current into rotational motion, which you can use to, say, propel a small toy car or whirl the propellers of a small drone. Off-the-shelf motors often spin too fast and too weakly for practical purposes, so they may require an attached gear train to slow down the rotation and give it more torque. A motor with a built-in gearbox is called a *gearmotor*.

SOLENOID

These helpful gadgets provide straight-line, rather than rotational, motion. They consist of a wire wound into a coil around a metal rod, which responds to current through the coil by thrusting out of (or withdrawing into) the housing. Solenoids are fast, dependable, and cheap, though not very strong or precise. They're good for valves and latches that should flip back and forth quickly.

STEPPER MOTOR

Basically a "digital motor," a stepper motor turns a small, precise fraction of an angle when fed a digital pulse, which allows exact control of the rotation (instead of turning freely when there's current). To turn it so many degrees, you just feed it so many pulses. Stepper motors are used in printers, disc drives, CNC tools, and many robots—they're perfect when you want to move a part to an exact position without checking it.

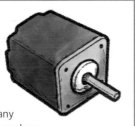

SPEAKER

You can use a speaker to convert alternating current into sound waves. There are several kinds, but the garden variety works a lot like a solenoid in reverse: A coil of wire carrying the current moves back and forth around a stationary magnet. The coil is attached to a diaphragm that transmits its vibrations to the air.

SERVO

This component doesn't usually turn in a continuous circle; instead, it rotates back and forth within a preset arc. The position of the shaft within its range is set by a control signal, which may be analog or digital, depending on the servo. Servos are often used in robots and RC vehicles.

PIEZOELECTRIC

Commonly used to make buzzers or beepers for high-pitched alarm tones, a piezoelectronic is a component that creates microscopic movements using the *piezoelectric effect*, a property of certain materials to physically expand when a voltage is applied across them.

MEET LIMOR FRIED

FOUNDER OF ADAFRUIT INDUSTRIES

Limor Fried—also known as Ladyada to her online followers—is one of the most well-known DIY inventors working today. Her open-source hardware company, Adafruit Industries, has shipped more than a million electronics products to like-minded creators around the world who use Fried's kits and parts to build everything from electronic demon costumes to robotic rovers.

Fried's fondness for electrical engineering and computer science traces back to an early love of video games. In her quest to digitally conquer, she frequently found herself "trying to think about how a computer thinks" and soon discovered that, with the shuffle of a few bits, she could pull off hacks such as giving herself unlimited lives on one of her favorite games, *Lode Runner*. "I was just amazed," she says of the experience. "Understanding computer science and electrical engineering is like becoming a magician."

Her invention career kicked off in 2003 with the creation of the SpokePOV, a bike wheel that lights up in customized displays. This creation also represented her first tinkering with microcontrollers—small, single-circuit computers that are at the heart of virtually all automatically controlled electronics. Originally envisioned for blinging out Fried's ride at Burning Man, the SpokePOV later became one of the first kits that she sold on Adafruit.

The company blossomed organically in 2005, while Fried was pursuing an electrical engineering degree at Massachusetts Institute of Technology. She had begun publishing instructions for various electronics projects on her website and was surprised by the response: She received several emails a day, all asking her for advice on how to get into DIY electronics. Even more surprising was the discovery that no adequate how-to guides existed that she could direct her readers to and

continued on next page

that easy-to-understand starter kits were equally nonexistent. Fried decided to fill that hole herself—and thus Adafruit was born.

She invested US$5,000 of her own money into prototyping and soon began manufacturing and selling parts and kits out of her apartment using 3D printers, molds, and laser cutters. Interest continued to build, and within a few years, Fried's one-woman operation expanded to a brick-and-mortar location in Manhattan, where she now oversees more than 75 employees.

During the last decade, Fried has designed around 300 Adafruit products herself, many of which were suggested by community members. Some of those creations originate from a weekly live video show called *Ask an Engineer*, in which Fried and guest experts answer questions, show off new designs, and solicit ideas. "I spend a lot of time looking at what makers are trying to do, and I keep all those 'problems to solve' in my head," she says. "Instead of me having to think of ideas, I get to tap into this huge ecosystem of creativity in the maker community." For example, cosplayers often used to complain to her that they couldn't easily add audio to their costumes. "One day I ran across a chip that—while not designed for cosplay prop audio—would do the job," she says. "Ta-da! A new design is born."

In addition to filling a niche for DIY electronics, Fried hopes Adafruit enables a diversity of makers to get involved in computer science and electrical engineering, regardless of background, age, or gender. It seems to be working. Recently, for example, a friend's 11-year-old daughter joined an *Ask an Engineer* discussion and posed a telling question: "Do boys do engineering, too?"

As Fried says, "She will never know a world where there aren't women doing engineering."

Q+A

Q: What tools are most useful for prototyping?

A: A really good multimeter and an excellent oscilloscope help me debug what's going on in the circuit. Also hand tools like a soldering iron. Electronics is a hands-on art, so having good tools really helps.

Q: Do you have an all-time favorite invention?

A: Yep, my favorite is the TV-B-Gone. A collaboration with Mitch Altman, this kit has it all: subversive pranking and learning microcontrollers, and it's a great project for beginners. It's a kit that, when soldered together, allows you to turn off almost any television within 150 feet (45 m) or more.

Q: Once you have an idea, how do you develop it?

A: It's pretty much just whatever I think is the solution. I'll look at what manufacturers have available and see what people have done before to get a baseline for improvement. Then I'll prototype, release it, and see how it's used. If it works well, I'm onto something. If not, I get a lot of feedback, then go back to the drawing board.

Q: Overall, what's the most difficult part of inventing, and how can an inventor overcome it?

A: Feeling like the thing you're trying to do has already been done, so what's the point? When you do something, you never step into the same river twice; by stepping in, you change the river. When we make something or we invent something, we change it just by doing it. Maybe it's more efficiently made, or has a different user interface or a different path . . .

Q: What are some great ways to market an idea?

A: The best advertising is good information. Explain to people that you've solved their problem in a cool way . . . They'll naturally head to you because they're looking for a solution anyway.

Limor's big breakthrough hit was the SpokePOV kit, an array of LEDs positioned along the spokes of a bicycle wheel so that, when the wheel spins, the lights appear in a pattern that the eye reads as a deliberate design. While the technology is neat, the trick is actually optical: Our eyes blend the rapidly moving lights into a single image. It's called *persistence of vision*—here's what Limor's device looks like at rest and in motion.

MAGNET

SENSOR

MICROPROCESSOR

UP TO
30 LEDS

BATTERY

With one circuit board mounted to your wheel, you'll need to hit 15 miles per hour (24 km/h) for the image to display. With the three shown here, you can cruise at 7 miles per hour (11 km/h).

The magnet mounted to the bike's fork triggers the sensor on the circuit board, which tells the microprocessor to make the LEDs blink in a pattern uploaded to the chip via a USB port.

052 MAKE YOUR INVENTION MOVE

The future is coming and it's full of robots. Robots—unlike smartphones, tablets, laptops, and most of the other devices we encounter daily—have to be able to move. From an inventor's perspective, this is a two-edged sword: On the one hand, movement opens up exciting possibilities as far as function goes. Machines that move can do lots of cool stuff that information-only machines can't. On the other hand, moving parts make things a lot less reliable—every latch, hinge, bearing, gear, and other interface between moving surfaces is an opportunity for dirt, grime, and wear to pile up over time and eventually gum up the works.

The more moving parts you add, the more of these failure points you introduce and the greater the likelihood that the whole system will break down because one bit has gone off the rails. For this reason, product designers usually try to minimize the number of moving parts required to make their invention do its thing, and you'd be wise to follow suit. And when you absolutely can't avoid mechanics, be sure to do them right—design and test carefully to ensure your shiny flagship product doesn't get a reputation for falling apart two months after you take it out of the box.

053 BONE UP ON BASIC MECHANICAL DESIGN

A lot goes into making stuff go. I can't wait to start building robots either, but truth be told that's pretty advanced kung fu. Especially if you're a newcomer to all this, you should practice treading water before jumping in at the deep end. Try familiarizing yourself with all that goes into making something move first.

PHYSICS

While engineers learn a lot of math in school, once out in the working world they tend to rely on intuition born of experience and software that simulates real-world problems realistically. But when it comes to the simple mechanics likely to be involved in consumer products, back-of-the-envelope math is simple and useful. So study how terms like force, torque, and friction translate to numbers and equations.

FASTENING

Unless your invention has only one part (like the doorstop—see #038), you'll have to figure out how the parts hold together. These choices will affect how durable your invention is, how easy it is to repair, and how much it costs to manufacture. Common fastening methods include adhesives (glue you stick between parts), welding (heat or chemical treatment that melts the parts together), and mechanical fasteners (screws, snaps, and rivets).

POWER

Any invention that does useful work requires a source of power. Nowadays, this is most often electrical, typically from a battery or wall plug. Batteries provide DC power and may be single-use or rechargeable, while a wall plug provides AC power that may be directly fed to the device's circuits—as in the case of a refrigerator—or may first be converted into DC for powering a smaller device.

ACTUATORS

Actuators turn electrical power into mechanical power. While there are many types (see #050), they tend to turn things in a circle (*rotary*) or move them back and forth in a line (*linear*). Some name *vibratory* as a third class—yes, the buzzer that makes your phone shake is an actuator too. Today, precision actuators usually take digital inputs, while actuators that don't require fine position control may still have analog control.

MATERIALS

Stuff! The particular substances that your invention is made of are incredibly important. The major classes are plastics, metals, and ceramics, though biological materials (such as wood) and composites (such as fiberglass) are also options. The semiconductors that make up microchips and other solid-state electronic components are another important class, but the odds are vanishingly small that you'll ever have to design such a device.

MECHANISMS

Once the actuators have got things moving, you may find that their movements need to be transformed. A toy tank, for instance, might need rotational movement to drive the treads but linear movement to sweep the turret back and forth. Enter mechanisms (see #037), which can make rotational movement linear and continuous movement intermittent or reciprocating, and transform fast, weak movement into slow, powerful movement (and vice versa).

054 BUILD A LAZIER SUSAN

You probably know the lazy Susan as a rotating platform in the center of a table that makes it easy to pass stuff around. I call this device a "lazier Susan" because it takes the same machine and adds a motor and some simple electronics so you can spin the platform just by flicking a switch. It's a handy thing to have in the house and a great way to wet your feet in basic electromechanics.

To build one, you'll need a drill and bits, screwdrivers, a pop riveter, a file, a marker, and soldering gear. You'll also need the following materials: a 16-inch- (40-cm-) round wood board, a 16-inch- (40-cm-) round cake pan, a lazy Susan bearing, wood screws, pop rivets, a DC gearmotor (I like the Solarbotics GM9), a motor mount or strap, a wheel, a four-D-cell battery pack and batteries, and a DPDT center-off momentary switch.

STEP 1 Drill for the bearing. Center the bearing on the bottom of the board, mark the mounting hole locations, and drill blind pilot holes for wood screws. Do the same thing on the bottom of the pan, this time drilling through clearance holes for the pop rivets. Now screw the bearing to the board.

STEP 2 Mount the motor. Cut a slot near one edge of the pan by drilling a line of holes and filing out the material between them. The slot should be just big enough that you can pass the wheel through it, like putting a coin in a vending machine. Fit the wheel onto the motor shaft, then position the motor inside the pan with the wheel protruding through the slot. Put the strap over the motor, mark and drill the holes, and secure with pop rivets.

STEP 3 Install the components. Position the switch in the side of the pan, clear of the motor. Drill a hole for the switch body and any mounting holes it might require, then secure it with the hardware that came with it. Position the battery pack in the center of the pan, underneath the bearing. Mark and drill the mounting holes, then rivet it in place

STEP 4 Wire the circuit. Solder the gearmotor leads to the terminals on one end of the switch, then solder a short set of jumper wires connecting these terminals with the terminals on the switch's other

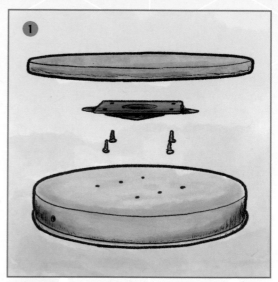

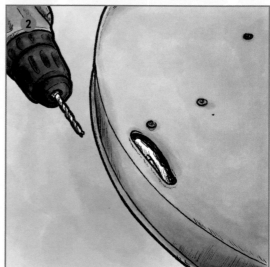

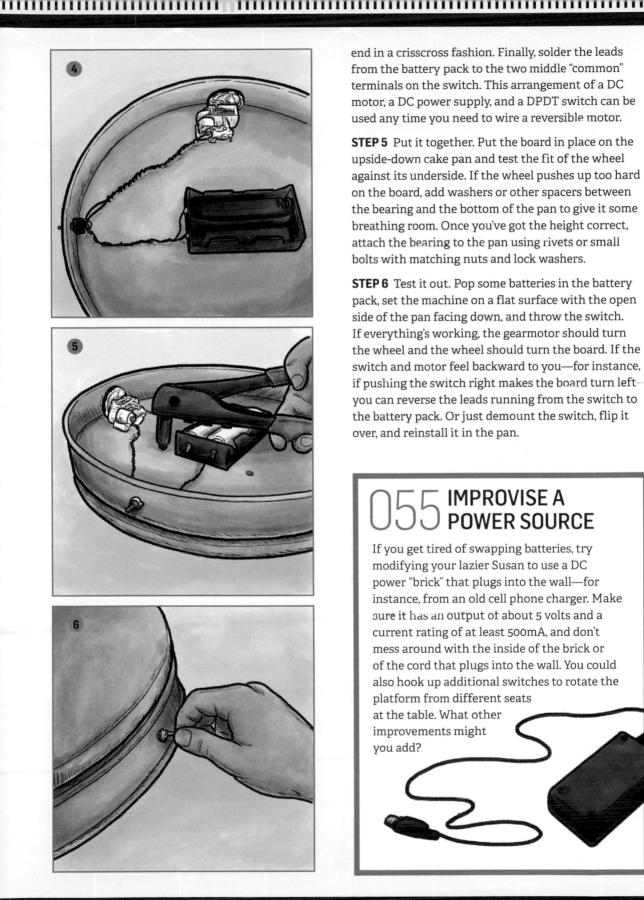

end in a crisscross fashion. Finally, solder the leads from the battery pack to the two middle "common" terminals on the switch. This arrangement of a DC motor, a DC power supply, and a DPDT switch can be used any time you need to wire a reversible motor.

STEP 5 Put it together. Put the board in place on the upside-down cake pan and test the fit of the wheel against its underside. If the wheel pushes up too hard on the board, add washers or other spacers between the bearing and the bottom of the pan to give it some breathing room. Once you've got the height correct, attach the bearing to the pan using rivets or small bolts with matching nuts and lock washers.

STEP 6 Test it out. Pop some batteries in the battery pack, set the machine on a flat surface with the open side of the pan facing down, and throw the switch. If everything's working, the gearmotor should turn the wheel and the wheel should turn the board. If the switch and motor feel backward to you—for instance, if pushing the switch right makes the board turn left—you can reverse the leads running from the switch to the battery pack. Or just demount the switch, flip it over, and reinstall it in the pan.

055 IMPROVISE A POWER SOURCE

If you get tired of swapping batteries, try modifying your lazier Susan to use a DC power "brick" that plugs into the wall—for instance, from an old cell phone charger. Make sure it has an output of about 5 volts and a current rating of at least 500mA, and don't mess around with the inside of the brick or of the cord that plugs into the wall. You could also hook up additional switches to rotate the platform from different seats at the table. What other improvements might you add?

056 GET PROTOTYPING TIPS FROM THE MASTERS

If there's one thing these prototyping gurus can all agree on, it's fail early and fail often. Here's what else they had to say.

"If you don't get a chance to fail, if you don't get a chance to try things and not get them right the first time and keep on doing it until you do get that specific kind of success, then you become so risk-averse that you in fact get an allergy to trying new things."
– *Adam Savage, maker and star of* Mythbusters

"I MADE 5,127 PROTOTYPES OF MY VACUUM BEFORE I GOT IT RIGHT. THERE WERE 5,126 FAILURES. BUT I LEARNED FROM EACH ONE. THAT'S HOW I CAME UP WITH A SOLUTION. SO I DON'T MIND FAILURE."
– James Dyson, inventor of the Dyson vacuum

"Our job as the game creators or developers—the programmers, artists, and whatnot—is that we have to kind of put ourselves in the user's shoes. We try to see what they're seeing, and then make it, and support what we think they might think. I don't think...[you] could create an experience that truly feels interactive if you don't have something to hold in your hand, if you don't have force feedback that you can feel from the controller."
– Shigeru Miyamoto, inventor of the Game Boy

"IF YOU ARE TRULY INNOVATING, YOU DON'T HAVE A PROTOTYPE YOU CAN REFER TO." *– Jonathan Ive, CDO of Apple Inc.*

"IF YOU'RE WORKING ON SOME PROJECT (MAYBE IT'S SOFTWARE, MAYBE IT'S HARDWARE), TRY TO THINK AT THE END ... 'HOW CAN I MAKE IT PERFECT? ALMOST BETTER THAN ANY OTHER HUMAN BEING WOULD? CAN I SHORTEN THE CODE? CAN I REDO IT WITH A LITTLE CLEANER STRUCTURE?' GIVE YOURSELF SOME OF THOSE REDOS."
– Steve Wozniak, cofounder of Apple Inc.

On prototyping showing you multiple potential uses: "After innumerable failures I finally uncovered the principle for which I was searching, and I was astounded at its simplicity. I was still more astounded to discover the principle I had revealed was not only beneficial in the construction of a mechanical hearing aid but it served as well as means of sending the sound of the voice over a wire." *– Alexander Graham Bell, inventor of the telephone*

"WE'RE MOVING INTO AN ERA WHERE IF YOU CAN THINK IT, YOU CAN MAKE IT. IF YOU CAN THINK IT UP THIS MORNING, YOU CAN PROTOTYPE IT THIS AFTERNOON." – MARK HATCH, FOUNDER OF TECHSHOP

"TESTING LEADS TO FAILURE, AND FAILURE LEADS TO UNDERSTANDING."
– Burt Rutan, engineer of the Voyager plane

On the proliferation of tools and services that aid in prototyping: "There are three big implications of this. First is that you don't need special skills to create things now—'I may have an idea but I don't know how to run a metal lathe, so that's the end of that.' Now you don't need to do that. Two, you don't need a lot of money to get a prototype out there. And three, you can order production at any scale, so you can start with 100 and then go to thousands or millions. So that was the structural change in the last four or five years. It's everything from 3D printers, CNC machines, laser cutters, and CAD software, to the cloud manufacturing services that exist from Shapeways to Alibaba." *– Chris Anderson, former editor of* WIRED *and cofounder of 3D Robotics*

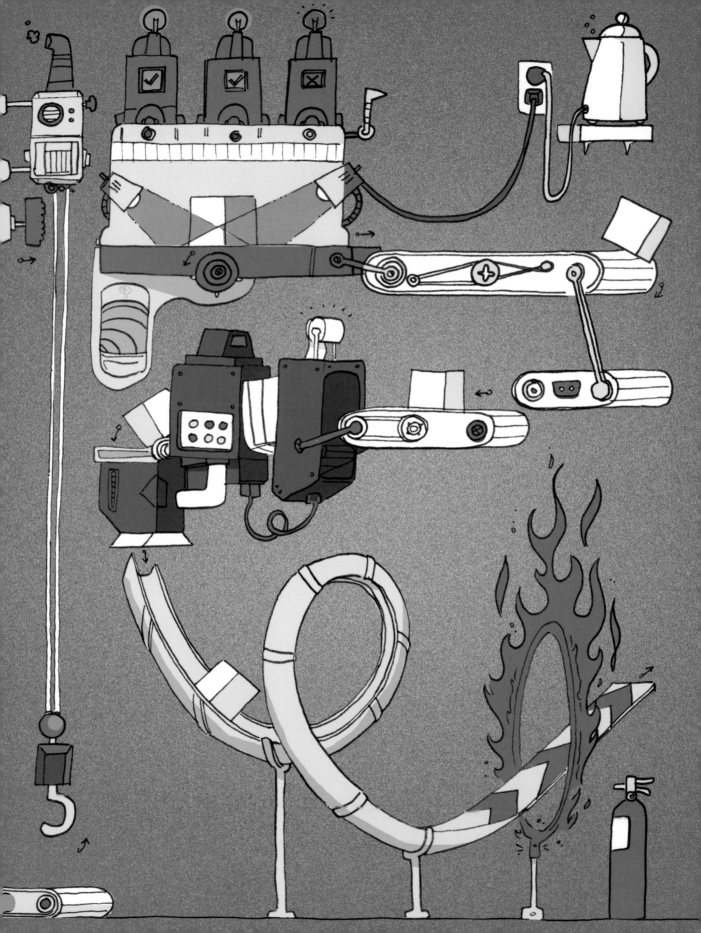

THE TESTING PROCESS

So you've got a prototype that makes you happy. It's time to invite loved ones and complete strangers alike to take a crack at using it and see if it makes *them* happy. Even if your invention is as simple as a paper clip, it's still vital to test it with potential real-world consumers as early and often as possible. Their feedback will be priceless—it'll help you improve your product and make it more likely to fly, not flop, out of the nest.

057 TEST IN THE REAL WORLD

The whole point of user testing is to observe potential customers interacting with your product, and then draw on their experiences to make it better. For your tests to have the most value, they should happen in a context that's as close to the real-world-use case as possible. So the kind of testing you perform will be very different if you're developing a smartphone app than if you're developing an outdoor power tool.

The amount of time, money, and other resources you can afford to spend is another major consideration. The more data you have, the more certain you can be about your conclusions, so test as much as possible.

Ideally, you'd test in a strictly scientific manner: hundreds of subjects, carefully trained administrators, artfully crafted questions, scrupulous data recording, and proper statistical analysis of the data. This is called *quantitative testing*. And you'd go through several rounds of testing so you could chart the improvement with each prototype. But most people aren't going to have those resources, so the question becomes: How do you make do? Answer: *qualitative testing*, which is a less rigorous but still rational approach to gathering, evaluating, and making decisions about data in the form of anecdotes, rather than hard numbers.

058 PAY FOR HARD SCIENCE

In the best-case scenario, you make business decisions based on rational, mathematical considerations of hard data, statistics, and expected value. And to make statistical decisions, you first need good statistics, which means data in the form of numbers—so-called *quantitative data*. And here's where it gets tricky for market researchers: Individual human beings do not, as a rule, think or speak in numbers.

You can interview a potential customer and record what she says, coax her to choose between option A and B, maybe even get her to give some answers on a scale of 1 to 10, but to extract cold, hard statistics from something as warm, fuzzy, and unpredictable as human interactions takes a lot of skilled work. You need a large

sample group, for starters, which means identifying, screening, scheduling, interviewing, and paying dozens or even hundreds of subjects, to say nothing of staff time spent collecting and analyzing the results.

If you can afford it, hire a market research firm. To up your bang-to-buck ratio, be clear on your objectives: Are you after feedback on your invention's usability, or do you want to know how your product compares to the competition? Help design a representative sample by sharing all you know about your prospective users.

If you can't afford quantitative testing, it's not the end of the world. Qualitative data, though more open to interpretation, can still help. And the best news is that you can gather it yourself.

059 CONDUCT QUALITATIVE TESTING

The practical alternative to formal scientific product testing is known as qualitative testing. Even if you're pursuing quantitative testing (see #058), you should do some of this good stuff too, which is all about getting to know your customers' experience in a more personal way.

STEP 1 Sit down with your potential customer, and—before even handing him the product—let him know that *he's* not being tested, the invention is. You want to make sure he feels comfortable, so try chatting about what you're up to and asking some friendly questions about his experiences with similar products.

STEP 2 Hand over the product and watch him use it, inviting him to think out loud about the process as he goes. It may help to give him specific tasks, if you're curious about how doable certain functions are. It's important to refrain from guiding him through these assignments—hang back and observe.

STEP 3 Document diligently. Take careful notes about your tester's experience. You can record it, but many people are uncomfortable being filmed or taped, and the test should be as low-pressure as possible. (If you do decide to record your sessions, get participants to sign a release.)

STEP 4 Ten solid interviews is a good goal for one round of qualitative testing. When those are in the can, review your notes and identify any recurring problems. Did four people struggle with the same component? Then you know it needs fixing.

060 HOST A FOCUS GROUP

A focus group is an arranged meeting of pre-selected people to have an informal discussion, administered by a trained moderator, about a particular topic. (Think group interview, but with a prototype in the middle of the table, and—if you're lucky—coffee and snacks.) The group usually has between six and eight participants, and they're encouraged to talk to each other more than to the moderator, who then tries to elicit a variety of opinions.

Focus groups go back to the 1920s work of pioneering American sociologist Emory S. Bogardus. Later, during World War II, researchers at Columbia University picked up the method to study the impact of propaganda on people's attitudes toward U.S. involvement in the conflict. Postwar, marketers and designers revived focus groups as a tool for getting qualitative feedback about a new product's appeal or usability.

Focus groups have a reputation for being a fast, cheap, and easy way to gather data, but they're not suitable for every purpose. The group response may be strongly influenced by the input of one or two outspoken individuals (hey, we've all been stuck in a room with *that person* before), and it's down to the moderator to create an environment where everyone feels comfortable sharing. Inventors should also recognize that, unless the real-world applications of the product involve a group context, the results may not reflect the experiences of lone users.

The best way to get a feel for what goes on in a focus group? Participate in one. If you've never done so, it's a worthwhile learning experience, and attendees are usually compensated. Market research firms in your area (and online) are always canvassing for potential focus group members, and if you keep your ears open, it's pretty easy to get selected when a group that fits your demographic profile comes around.

061 FIND SOME GOOD TESTEES

Ahem. Well, you've got to look in the right place. Obviously, you want testers who match your customer profile: people who have a need your product can fulfill and who are willing to pay what you expect it to cost. How do you reach them? Social media is very powerful for this purpose.

If you've got a company social media presence already, you can start by posting a "product testers wanted" ad there, stating briefly who you're looking for and mentioning up front what you're paying in compensation. Oh, yes: You need to pay. How much varies depending on your customer profile, but people should feel fairly rewarded for their time and energy. Between US$75 and US$125 per person is a good rule of thumb for general consumer market products.

In your post to your social media accounts, ask people to tag or share with anyone they think might be interested. You can also try posting public ads on websites or even in print newspapers, or take the good old-fashioned "press the flesh" route, whereby you approach strangers in public places like malls, university common areas, or coffee shops. If you can find a conference or other event where your customers congregate, your time and energy will likely be better spent than in a generic public place—you'll spend more time talking to suitable candidates and less time weeding out unsuitable ones.

Once you identify interested people, direct them to a screening questionnaire (see #062 at right). The easiest way is with an online survey tool. Follow up with everyone who responds and thank them personally; those who meet your profile should be scheduled for a testing appointment.

062 MASTER THE QUESTIONNAIRE

Questionnaires are a great tool for IDing members of your target market—aka, potential users who should test your product. If you've invented, say, a blood-testing device for people with type 2 diabetes, you wouldn't want to solicit feedback from college athletes who've never known anyone with that condition. Here's how to write a questionnaire that effectively screens for subjects who fit your customer profile.

CAPTURE THE BASICS Start with a section on gender, age, race/ethnicity, location, education, employment, and income bracket, as it helps to know as much about your testers' backgrounds as possible.

CHOOSE A FORMAT There are two flavors of questions: *open-ended* queries that prompt respondents to describe their experiences in their own words, and *closed* questions that present them with choices. A closed question may ask a person to rank a statement's importance or her level of agreement with it, or it may ask her to choose a level of likelihood that she'll do something in the future. These questions are generally easy for people to answer quickly, and they allow for easy data calculations, but it helps to have a mix of both.

WRITE SIMPLY This can't be overstated. Writing a questionnaire is not the time for your inner Bukowski to emerge. Keep your language plain, sentences short, and meaning clear.

DOUBLE UP We humans occasionally answer questions the way we think we're supposed to rather than the way we actually feel. To skirt this bias, draft three or four slightly different versions of your core questions, then average the various responses to get close to the real answer.

063 CRAFT YOUR USER STORIES

A *user story* is exactly what it sounds like: a short narrative written from the perspective of an imagined user that describes how that person works with your invention to achieve a goal. If your product were, say, a camera-equipped flying drone that can automatically follow you around and film your actions, one user story might go something like this:

"As a professional surfer, I want to film myself riding waves so I can go back later, analyze my technique, and improve. I should be able to carry the camera on my board while I'm paddling out, toss it up in the air before I catch a wave, and then be able to easily replay what I just did once I get to shore."

Writing a user story—or several, in the likely case that your product will serve many needs—helps you articulate your hypotheses about how customers might experience your invention. Resist the urge to put off writing these narratives until after testing: Inventors are usually surprised by what inexperienced users do when they first encounter the prototype, and comparing the testing data with prewritten user stories is a great way to understand where your blind spots are.

As for what's lurking in those blind spots, be open to pleasant surprises. You may hear uses for your product that you'd never dreamed of—perhaps a wedding photographer who programs the drone to follow the bride around the dance floor, or a farmer who uses it to locate his herd on an expansive ranch. The more uses testers come up with, the greater your potential for market share.

EXERCISE

064 DISSECT A COMPETING PRODUCT

If you're competing against physical products, go buy (at least!) one of them. The ideal would be to build a complete physical library of all competing products for you to reference throughout the development process. Start by dissecting them one by one.

STEP 1 Make a note of where you bought it, what it cost, how it was merchandised in the store, and if other people seemed to be buying it. If there's prominent real estate devoted to similar products, then you at least know consumers are into the concept and retailers know how to position it on the floor.

STEP 2 Take it home and unbox it, paying careful attention to the packaging. (How successful is its messaging? Do you immediately know what the product does? What sort of promises does it make to the consumer?) Then, without throwing anything away, start learning to use the product.

STEP 3 Once you've got the hang of it, make a guess about what it looks like inside. How do you think your competitors made it work? Then it's time for the teardown: Pop out the battery, bust out your screwdrivers, open up the case, and see how close your guess was. Try to proceed nondestructively, with the goal of someday putting Humpty Dumpty back together again, but if you have to break stuff to get in there and understand, do it. Take pictures as you go, and don't stop until all the internal parts are laid bare.

STEP 4 Make a complete parts list, using a spreadsheet to keep track of individual pieces, their identifying numbers (if any), and how many of each there are.

STEP 5 When you're done, there's no need to put the thing back together right away. Store the parts, packaging, and everything that came with it in a sturdy box in a secure location—you may want to refer to specific inner workings to learn how they solved certain problems later. If you can afford to, you might even buy a second working model to keep on hand.

065 DON'T GET CAUGHT IN THE IRIDIUM TRAP

If you're old enough to remember the years before everyone had a cell phone, you're likely old enough to remember the Iridium Network. Named for the chemical element iridium and its 77 electrons, the 70-odd satellites of the Iridium constellation began launching into Earth's orbit in the early 1990s. By the time the company deployed its globe-spanning, satellite-based cell phone service in 1998, the total cost had run to an estimated US$5 billion.

Despite all the investment in the project, it was a disaster. The phones were bulky, expensive, and unreliable compared to ground-based cell services. You had to pay thousands for a phone that weighed a pound, frequently dropped calls, didn't work over the regular cell network, and couldn't be used indoors. And for that privilege you paid US$5 a minute to talk. True, you could place a call from an oasis in the Sahara to a beach in Thailand, but the number of customers who were ready to pay that much

to do so didn't even come close to fulfilling the company's expectations. Or its obligations.

Iridium Satellite LLC filed for Chapter 11 bankruptcy less than a year after launch and, in 2001, the satellites and remaining assets were sold off for a scant US$25 million. There was serious talk of de-orbiting it to prevent it from becoming hazardous space junk.

How did it go wrong? With so many ducats on the line, Iridium of course did market testing—it screened hundreds of thousands of potential subjects and tested more than 20,000 people. The problem was methodology. Survey questions spoke of "a small handset that fits in your pocket" that sold for "a reasonable price" with service "not limited like a cell phone," pitching a dream far removed from the product coming down the pipe. The engineers were too wrapped up in the feat of creating the satellite system to make the product that they'd promised—and tested with—consumers.

HOT END

EXTRUDER

PRINTED
MODEL

PLASTIC
FILAMENT

066 MEET THE 3D PRINTER

Any solid shape can be thought of, abstractly, as a stack of flat, two-dimensional slices. This will be a familiar idea to anyone who's ever studied calculus. (Don't worry, that's optional.) For our purposes, it's enough to understand that the whole idea of 3D printing is based on this fundamental fact. And though there is a surprisingly large number of different machines sitting around with labels reading "3D printer," they all work in fundamentally the same way: The solid shape you want to print is first sliced up into flat cross-sections, in software, and then the machine works through these layers one at a time, from one side to the other, building up the form as it goes.

Of course, this process can get complicated—and quick. In particular, with some printer types, shapes with overhanging features will often require printing temporary support structures to keep them from falling or sagging before printing is done. Software will often handle such technical factors automatically, but not always.

But stick with it, and you'll be well rewarded. 3D-printing is a heckuva lot of fun, and the ability to rapidly print, test, and tweak parts for your prototype can pay off big in calendar time. It's generally worth the tinkering and—real talk—occasional cussing these machines sometimes require. Let's dive in.

067 PICK THE BEST PRINTING METHOD

There are many types of 3D-printing processes; ultimately, the choice comes down to material, required precision, and cost. Also, some processes have quirks you may need to consider—for example, if you print a closed hollow box with a liquid-vat process, there will be no way to remove the feedstock, and you'll end up with a box full of liquid. Other factors—like minimum feature size (aka resolution) and the need to print support structures—can be important too.

POWDER BED FUSION

HOW IT WORKS A bed or tank of very fine powder is fused together, layer by layer, by a laser or other heat source that moves according to your design. After each layer fuses to the lower one, the tank drops down, a fresh layer of powder goes on top, and the cycle repeats.

MATERIALS Rigid and elastic opaque plastics, aluminized plastic, aluminum, stainless steel

USE IT FOR Most general situations and materials

MATERIAL JETTING

HOW IT WORKS An inkjet head directly applies liquid to a build surface, which is often heated to aid adhesion or curing. The liquid is usually a photopolymerizing plastic, in which case each layer is exposed to an ultraviolet light to harden it after deposition.

MATERIALS Various plastics and elastomers, transparent and opaque

USE IT FOR Prints that require more than one material

PHOTOPOLYMERIZATION

HOW IT WORKS In this method, the 3D-printed object emerges from a vat of light-reactive liquid plastic, layers of which are selectively hardened and fused into an object, one at a time, by a bright light. In *stereolithography* (SLA), the light source is a scanning laser beam, but some newer methods swap the laser beam with what is essentially a digital video projector.

MATERIALS Opaque and translucent acrylic plastics

USE IT FOR Printing extraordinarily fine details

EXTRUSION

HOW IT WORKS A coiled filament is fed into a hot nozzle, which pushes it out in a thin bead of molten material that bonds to earlier deposits. Layers are filled in with these fine lines. Known as *fused-deposition modeling* (FDM) or *fused filament fabrication* (FFF), extrusion is the cheapest, most common process.

MATERIALS Plastic or wax

USE IT FOR Cheap prints at home on your own equipment

BINDER JETTING

HOW IT WORKS As in powder bed fusion, the model slowly grows into a tank of very fine powder. However, instead of using heat to fuse the particles, an inkjet head applies water, glue, or another chemical binder to join the layers. Bonus: The use of colored binders allows for full-color 3D prints.

MATERIALS Gypsum "sandstone" and edibles

USE IT FOR Printing in full color

068 BUY OR BORROW A 3D PRINTER

Do you need to purchase a 3D printer? Short answer: No. Don't get me wrong—it's definitely a fun thing to have. And from a more practical perspective, the ability to rapidly evolve designs by quickly repeating the print-test-revise cycle can be very powerful, and that process can be stunted if you don't enjoy exclusive access to the necessary equipment. That said, there are lots of 3D-printing contractors in the world, and most of them have online portals that allow you to upload 3D-software models and receive a physical part, by mail, in a relatively short time.

069 PRINT YOUR FIRST 3D MODEL

Fans of *Star Trek* need no introduction to the idea of the replicator. While we're still a long way from "Tea, Earl Grey, hot," it's relatively cheap and easy to have a machine materialize a solid part, in whatever shape you want, from a wide variety of bulk raw materials. And though there are many different feedstocks, processes, and machines, at the big-picture level, the workflow for replicating objects goes like this.

STEP 1 Make a digital model. Just as you have to write a document file to print a term paper, you have to make a 3D-model file to print an object. To do this, you'll need a program that lets you write 3D models. There is a plethora of these programs available but SketchUp and OpenSCAD are user friendly and free, or nearly free, online (see #021–024).

STEP 2 Preprocess your model. Once you have a 3D model you like, you'll need to prep the file for printing. The amount of work required depends on the printer, but by and large there are three things to consider.

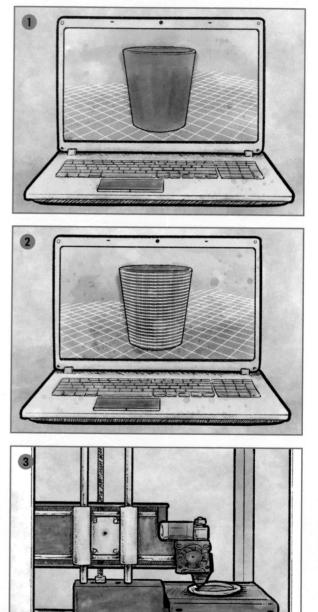

- Your model should be properly *manifold* or watertight. This technical term indicates that your model contains no errors that would make it impossible for the printer to tell the inside from the outside. Take a simple cube shape, for example. If you leave off one corner (or, as is more common, accidentally have two corners defined in the same place or nearly the same place), it stops being a solid and becomes a mess of flat planes, each of which has no thickness! The printer will understandably give up if you tell it to produce an impossible object.

- Your model should be in the proper file format. For most home printers, this will be an *STL file*, which is an acronym for stereolithography. Even if you're not using the stereolithography process, most printers still expect this file type. Your 3D-modeling software will likely let you export STL files; if not, there are programs that convert most 3D-model formats.

- On older, cheaper, or open-source models, you may need to slice the file before printing. In this step, a program called a *CAM processor* turns the 3D-model file into a set of instructions for the 3D printer's mechanical parts.

STEP 3 Print the model. With most common desktop fused-deposition modeling (FDM) equipment, this means first loading the filament, maybe preheating

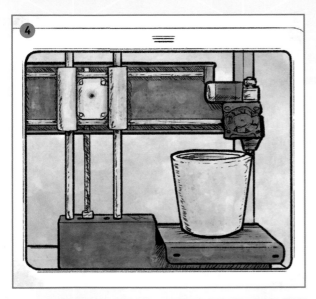

the nozzle or build platform, and even manually setting the print head in its starting position.

STEP 4 Wait it out. The print job may take minutes or hours. The printer control interface—whether it's a control panel on the machine or a virtual window on a computer—should show your wait time. For your first few prints, stay close to the printer so you'll know early on if something goes wrong.

STEP 5 Postprocess your print. If it's a robust, solid shape, you can likely just grab it and pull it from the build platform. If it's a lacy, delicate thing, you may need a hobby knife to get it clear without breaking anything. Some machines print the model on a *raft*— a loose, plastic mesh that steadies the model during printing—but you can easily remove it with your fingers or a hand tool too. Ditto with any temporary structures printed to support overhanging parts.

STEP 6 Congratulations! Finishing your first 3D print is an event worth celebrating. The model shown here is a shot glass first uploaded to RepRap.org by Sebastien Bailard on July 2, 2008. Back when the only way to get a home 3D printer was to build it yourself, this model became the traditional first test print for anyone who had finished building a RepRap 3D printer. If your printer is functioning properly, the completed shot glass will be literally watertight. It can also hold liquids more festive than water if you're of age and inclined to toast the moment. Bottoms up!

070 HUNT FOR 3D MODELS ONLINE

Making the 3D model is probably the hardest part of this process. The good news? Somebody else may have already done it for you. There are lots of 3D-model-sharing websites out there that you can search—Thingiverse and GrabCAD are two leaders in the field. Even if nobody has designed exactly the thing you're after, it may be possible for you to download somebody else's model and modify it to work, which will likely be easier than starting a new model from scratch.

MEET MEG CRANE
INVENTOR OF THE HOME PREGNANCY TEST

The year was 1967—the year the U.S. Supreme Court ruled that laws prohibiting interracial marriage were unconstitutional, the British Parliament decriminalized same-sex acts and legalized abortion, and the Beatles released *Sgt. Pepper's Lonely Hearts Club Band*. And the year that Meg Crane, a 26-year-old product designer working for a pharmaceutical company in New Jersey, forever changed women's access to information about their reproductive health by inventing the home pregnancy test.

Crane had recently been hired at Organon Pharmaceuticals to help them launch a line of makeup. "I'd freelanced at *Harper's Bazaar*, so I'd studied fashion and illustration—I was attuned to the world of cosmetics and was a good candidate for the job," she says. Most of her days were spent sketching promotions for lipsticks and lotions, but one day she noticed a rack of test tubes in the office. "A lab technician explained that they were pregnancy tests, and that if a woman thought she might be pregnant, she would go to her doctor and give a urine sample. Then he would send it to the company for testing," Crane recalls.

The test itself didn't look like much—just a test tube suspended over an angled mirror. The lab would mix the urine sample with a chemical in the test tube, then wait two hours for the results to become apparent via chemical reaction. The tell-tale sign of pregnancy? A red dot at the bottom of the tube, which you'd only be able to see in the mirror because—if you picked up the test tube to look at the bottom—you'd likely jiggle it, causing the reaction to disperse. Crane was gobsmacked: "I immediately thought that it was so simple a woman should be able to do it herself."

Crane was also inspired. She started spending her evenings on a consumer prototype, housing it in a clear plastic paper-clip box from a Japanese stationary store. "I cut some holes in the plastic to

continued on next page

hold a test tube and an eyedropper. And I used a piece of Mylar, angled at the base of this little box, as a mirror," she says.

In many ways, the prototype proved to be the easiest part of the project. Crane's male bosses at Organon weren't enthusiastic, fearing that a home pregnancy test would siphon dollars away from doctors, the company's main source of income. Plus, they weren't convinced that women could accurately use it: "They thought women would be too emotional," Crane recollects. "The other concern was that it was not a moral thing to put on the market. I was approached by people saying I was . . . not exactly the devil, but close to it." Luckily, cooler heads prevailed, and Organon hired a team of product designers to come up with a concept. Crane was not invited to share her prototype, but she crashed the meeting anyway. Her transparent, elegant design beat out a table full of frilly tests, all embellished with florals, tassles, and ribbons. "My design was the most scientific and sturdy. Nothing else was needed—except for a nice quiet shelf for a woman to hover over for two hours," she jokes.

Organon tested Crane's prototype in New York City hospitals, successfully confirming that women would be able to put it to proper use, and then wanted assurance that the market wouldn't revolt in those pre-Roe-vs.-Wade days. Meanwhile, Crane and her partner—an ad man she met when he advocated for her design in that fateful concept-review meeting—had started their own firm, and Organon hired the couple to run a market test in Canada. "I interviewed friends and family about hearing that they were pregnant from their doctors," Crane says of this stage. "They kept saying how hard it was to wait weeks for the results." Crane also learned what they wanted in the packaging: clean graphics, a calm blue color, and discreet branding.

There was some outcry from religious groups, but after a year it was clear that the test sold. Some tweaks were made—the chemistry was refined, and a vial of distilled water was added so the urine sample could be reliably diluted. And finally, after a long wait for FDA approval, Crane got to see it hit U.S. stores in 1977.

Not long after, she was at a party where a friend happily announced the results of a test she'd taken that day. "Another friend said to the group, 'Did you know Meg invented that?' People said no way," Crane reminisces. "But it was wonderful to hear women share their news and know they'd learned it from the test."

Q+A

Q: What's your favorite tool?

A: One of my favorite things to work with is a particular pencil, a Blackwing 602. It comes from Germany and it's the best pencil for doing the *New York Times* crossword puzzle.

Q: What's your earliest memory of tinkering?

A: I tried to make an umbrella out of wax paper and some sticks when I was a kid. Not very effective, as you might imagine.

Q: What inventors/makers inspire you?

A: Alan Turing. His story is so heartbreaking and amazing to me.

Q: What was your greatest challenge in making your invention happen?

A: The initial negativity of the company about my idea. It's important to have people on your side who will stick with you—who have some influence and can help you get it through.

Q: Do you ever get stuck when you're working on an invention? What do you do to hit refresh?

A: Try turning it upside-down—that sometimes does the trick. And approaching it from someone else's point of view can prove interesting. My partner would come up with design ideas that I didn't think could work at all at first. We learned a whole lot by collaborating.

Q: Do you have any words of wisdom for aspiring inventors?

A: You don't know where you're going to wind up—you don't have to get stuck. If something else comes along that you're interested in, jump to it. Also, I signed away my rights to Organon for US$1. Today I tell people that if they are in that situation, they should walk away, or come back with a lawyer. I never did see that dollar.

Over the years, the home pregnancy test has evolved from the boxed version you see here to a wand (or pee stick, to use the parlance of our times). Crane's first design succeeded because it provided consumers with an honest view of the science at work. It also didn't pander to gender stereotypes with decoration or the usual Pepto-Bismol-inspired color palette. Still, it took years of testing to convince pharmaceutical companies that women could and should use the product. Here's how the simple science and easy-to-use design worked.

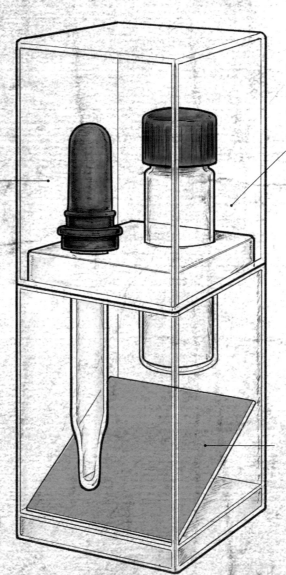

The test tube holds chemicals that detect high levels of human chorionic gonadotropin (hCG) in the urine sample, indicating a pregnancy. If a sample was positive, a red dot would form at the bottom of the tube.

An eyedropper allowed women to easily transfer urine samples to the test tube.

A mirror at the bottom reflects the bottom of the test tube up to the consumer so they don't have to lift the tube and risk jarring the contents.

The entire test is encased in clear plastic so light could illuminate the results and consumers could understand how it functioned at a glance.

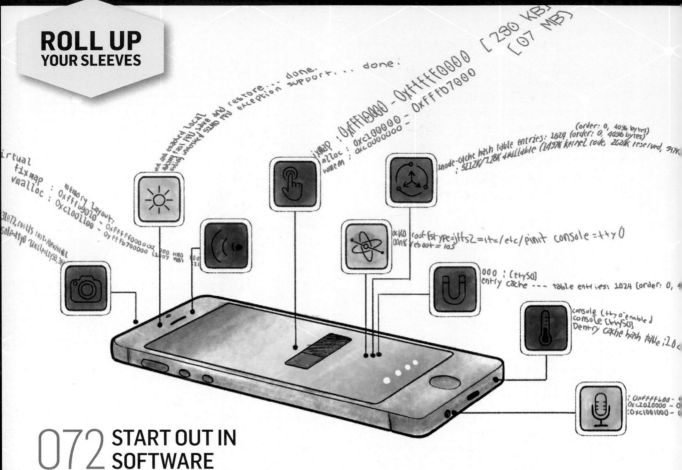

072 START OUT IN SOFTWARE

The smartphone in your pocket is *hardware*; the apps that run on it are *software*. This distinction is pretty intuitive to most people nowadays, but it's worth underlining here. While we still call them phones, a smartphone is a computer, and, as of this writing, any run-of-the-mill, off-the-shelf model incorporates at least eight sensors that allow it to receive information from the environment. At a bare minimum, it'll have an onboard *accelerometer* (a device that measures acceleration), a microphone, a *gyroscope* (which calculates orientation and rotation), a *magnetometer* (a digital compass), a proximity sensor, a light sensor, a touch-sensitive display, and a thermometer. That's not counting the two cameras (front and rear) and the three radios (cellular, Wi-Fi, and Bluetooth).

What the computer does with all that information is down to software, which is intangible and abstract. Software is a set of instructions, called a *program* or an *application*, written in a language the computer can understand, called *code*, which tells it what to do and when to do it (see #075–076). The following metaphor should be taken with a boulder of salt, but

in biological terms: Your brain is hardware, while the things you learn to do with it are software.

Writing software is much easier than building hardware, which is largely the point: Instead of building one device to make phone calls, another to take pictures, another to play music, another to take dictation, and so on, we build one very small computer with all the input and output components we can cram in, and then we let the software tell it how to be a phone, a camera, an MP3 player, a recorder, or anything else we can program it to be. As a business model, software has its own challenges, but the barrier to entry is much lower than in hardware for the simple reason that you never have to *physically* build anything. So if you're starting from scratch, and if what you want to do can be done by writing software to run on somebody else's hardware, it's worth trying it that way first before jumping to manufacture your own device.

But before making these decisions, let's learn a bit about how different software systems work and how—of course—you can get started coding one.

073 GET TO KNOW EMBEDDED SYSTEMS

Your phone isn't the only widget that's a computer in disguise. If your car was made after 1980, it's likely one too. As is the plane you fly in to go on vacation, the traffic light you stop at on the way to the hotel, and maybe the hotel itself. If the Internet of Things hype pans out, almost every device you own (refrigerator, hair dryer, light bulb) will soon have a computer inside.

A computer that's permanently built into a larger device is called an *embedded system*. As the price of programmable microchips has fallen, engineers increasingly opt for embedded systems over task-specific circuits, as it's cheaper to solve problems by reprogramming a chip than by redesigning a board. Here are some common types of embedded systems.

MICROPROCESSOR

This gizmo is a computer central processing unit (CPU) on a single microchip. It can be programmed for many uses and runs through its operations on a set frequency—say, 3 gigahertz (3,000,000,000 operations per second) or, in the early days of desktop computers, 4.77 megahertz (4,770,000 operations per second).

MICROCONTROLLER

A CPU with digital memory and input and output circuits all on a single chip. The memory will be either *nonvolatile*, which can be programmed and will remain intact when the circuit's power is turned off, or *volatile*, which goes away as soon as you nix the power. (It's used by running programs to temporarily store data during processing.)

OPERATING SYSTEM

This software runs whenever a computer is on, and it manages how different applications (think web browser or word processor) use the hardware. When choosing an operating system for an embedded system, ask yourself: What chip(s) is it going to run on? And after that: Does the job require real-time computing?

REAL-TIME COMPUTING

What computers do, in general, is take a bunch of input, process it, and then deliver some output or result. Sometimes, these results need to happen fast, so the electronics must be designed for *real-time computing*: an approach to hardware and software that mathematically guarantees output within a certain period of time.

074 TRY OUT A DEVELOPMENT BOARD

Companies that make integrated circuits want you to build prototypes using their chips. Their logic is solid: Once you've got everything working in a prototype, you won't want to rock the boat when you shift to manufacturing, so you'll likely choose to mass-produce using the same components. That means you'll buy tens of thousands of their chips. *Cha-ching*!

While sneaky, these companies usually provide lots of free or at-cost resources to help you understand how their chips can help you. The most useful is probably the *development board*, a circuit board designed for experimenting and prototyping, which mounts a particular chip or set of chips in a way that lets you easily wire it to a breadboard or other components on a test bench. It may come with user-friendly programming software as well. Once you've decided you're interested in a particular microchip for your product, the first thing you want to do is order a development board, build a prototype with it, and see how you like the results.

075 SPEAK YOUR COMPUTER'S LANGUAGE

To program a computer, you break a task into smaller and smaller steps until the actions are simple enough for a computer to do. Then you explain those actions in a computer's language. There are dozens of popular programming languages out there, but they're not full languages like Mandarin or Swahili. They're simple English with very special rules. The rules are different for each language, but they have common concepts.

SPECIAL PUNCTUATION In programming, forgetting a parenthesis or using a comma instead of a period can stop a program dead. Here are some common examples of special punctuation.

COMMENTS These are used as a chance to inject some plain English into your code and are ignored by the computer. They often start with # or // and end at the end of the line, or begin with a /* and end with */.

CODE BLOCKS Handy for grouping a bunch of code together as a single unit. This is usually done with curly brackets, but some languages group code by how far the line is indented.

END OF LINE This is commonly indicated by a semicolon, while some languages are smart enough that you can just hit return.

VARIABLES The labeled boxes of coding. You can put a single thing in the box, for example, a number or word. Then you can do things with the boxes, like add or compare their contents. And you can give them descriptive names so you can tell what kind of thing they're supposed to hold.

CONDITIONALS These let the computer make decisions by comparing things, usually variables.

Here's an example:

```
IF (X > 10) {
    // DO SOMETHING.
} ELSE {
    // DO SOMETHING ELSE.
}
```

In plain English, this reads "If the value of variable X is greater than ten, then do whatever is in the first block of code. Otherwise, do whatever is in the second block of code." Conditionals sometimes use borrowed and improvised logical symbols, like != for "not equal to" or >= for "greater than or equal to."

LOOPS Use a loop to make computers do the same thing over and over without complaining. They're like conditionals with something extra going on:

```
FOR (X=0; X < 10; X=X+1) {
    // DO SOMETHING.
}
```

In plain English: "Set X to zero. If X is less than ten, do the block of code. When the block is complete, add one to X. If X is less than ten, do the block of code..." and so on, adding one to X each time through. When X is no longer less than ten, the loop stops.

FUNCTIONS You can split your oft-used code into a function that you can reuse without copying every time. Here's an example in the JavaScript language:

```
// A SMALL FUNCTION TO RETURN THE
CUBE OF A NUMBER.
FUNCTION CUBE(NUM){
    VAR OUTPUT = NUM * NUM * NUM;
    RETURN OUTPUT;
}
ALERT("THE CUBE OF 12 IS " + CUBE(12));
// 12 CUBED IS 1728
ALERT("THE CUBE OF 42 IS " + CUBE(42));
// 42 CUBED IS 74088
```

Here, we send a number (num) to the function, which cubes the number, returns the answer (output), and displays it on-screen. You can find a basic JavaScript program that combines these concepts online.

076 DECODE DIFFERENT PROGRAMMING LANGUAGES

Which coding language should you learn? Depends on what you want to do. While it seems like there are an infinite variety of programming languages, these eight are among the most popular right now.

C

Known for its speed, stability, and cross-platform appeal, C is the oldest and most commonly used language. It's referred to as low-level, meaning it's quite malleable and not very specialized. It's also procedural, so the code instructs everything it does and is easy to read.

SAMPLE USES Operating systems, embedded system applications, end-user apps

WHO USES IT Everyone, by proxy

C++

Built on the foundation of C, C++ is a general-purpose, intermediate, object-oriented language (in which you make a world of objects and assign them behaviors). Changing one thing could affect others—tricky for big teams coding simultaneously.

SAMPLE USES Software driven by graphical user interfaces, performance-critical applications

WHO USES IT Adobe, Microsoft, Autodesk

JAVA

Deriving much of its language from C and C++, Java is a class-based, object-oriented language (meaning it divides objects into classes and gives them roles). It's modular—so hundreds of engineers can work on a project together simultaneously—and multithreaded, so one program can do many things at once.

SAMPLE USES Client-server web applications, enterprise software, games, mobile apps

WHO USES IT Android, Netflix, Amazon

JAVASCRIPT

One of the three core technologies of the World Wide Web, JavaScript is high-level and designed to add effects to webpages and run in a browser.

SAMPLE USES Interactive or animated web functions, game development, desktop applications

WHO USES IT The Internet, all major web browsers

PYTHON

Python is a general-purpose, high-level scripting language, meaning it's very pragmatic for high-level, complicated tasks. If Python had a motto, it would be "Whatever it takes to get the job done."

SAMPLE USES Web applications, software products, scientific computing, video games

WHO USES IT Google, Firefox, NASA, Yahoo

PHP

PHP is an embedded language, so it can be directly embedded into an HTML source document. It uses scripts stored on a web server to customize a website's response to requests.

SAMPLE USES Dynamic websites and apps, web content management systems

WHO USES IT Facebook, WordPress, Digg

SQL

An acronym for *structured query language*, SQL is used to manage data in relational database-management systems. It organizes the data much like an Excel spreadsheet, so it's easy to search.

SAMPLE USES Mobile apps, databases for businesses, hospitals, universities

WHO USES IT Google, Skype, Dropbox

C#

Originally developed by Microsoft for Windows, C# ("C sharp") is a general-purpose, object-oriented language that is built on C and C++ and supports multiple programming models.

SAMPLE USES Applications that run on Microsoft's .NET Framework, business applications targeting Windows environments, game programming

WHO USES IT Microsoft, Stack Overflow, Dell

077 PROGRAM YOUR FIRST BOARD

Let's program a very simple computer—the type that might be used in an embedded system—to do something easy: Blink an LED on and off at a variable rate. We'll use an Arduino Uno, a development board produced by Arduino LLC, which publishes open-source hardware and software, and makes money by licensing to companies who manufacture and sell boards using its designs. (See #139 for our conversation with one of its founders.) If you've never programmed a piece of hardware before, Arduino is the best platform to get you started.

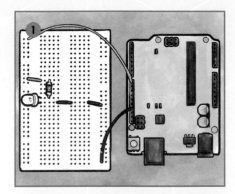

You'll need a personal computer and a USB A/B cable, plus an Arduino Uno board, a solderless breadboard, a red LED, a 120Ω resistor, red and black jumper wires, and an optional three-AA battery pack with 5.5/2.1mm center-positive barrel plug.

STEP 1 Hook up the components. Connect the resistor, LED, breadboard, and Arduino as shown. The red wire goes to the Arduino's pin number 8, while the black wire goes to its GND pin (which stands for—you guessed it—"ground").

STEP 2 Connect to your computer. Visit Arduino.cc and download the Arduino software. Follow the instructions to install it, then connect the USB cable between the Arduino and a free USB port on your computer. Small green and orange lights on the Arduino board should come on, and you'll likely receive some kind of notice on your computer screen that it's connected.

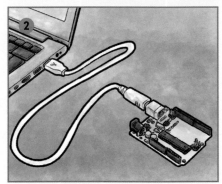

STEP 3 To enter the code, first launch the Arduino application. Arduino programs are called *sketches*, and each one consists of two sections: *setup* and *loop*. The stuff in the setup section comes first and only gets run once at the very beginning. It's generally used to, well, set up how the board is going to work and other factors that aren't going to change as the sketch runs. The stuff in the loop section runs over and over again as long as the Arduino is powered on. Enter the code shown here and save the file.

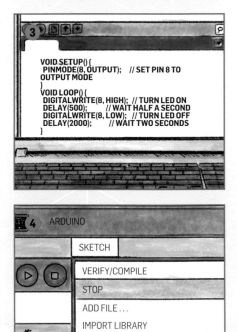

```
VOID SETUP() {
  PINMODE(8, OUTPUT);    // SET PIN 8 TO
OUTPUT MODE
}
VOID LOOP() {
  DIGITALWRITE(8, HIGH);  // TURN LED ON
  DELAY(500);             // WAIT HALF A SECOND
  DIGITALWRITE(8, LOW);   // TURN LED OFF
  DELAY(2000);            // WAIT TWO SECONDS
}
```

STEP 4 Program the Arduino's microchip and test it. Select Verify/Compile under the Sketch menu. You should get a message that reads, "Done compiling," which means you entered the code without mistakes. If you don't get this message, check to see if you typed everything correctly. (The stuff after the double slashes can be typed in or not; these are called *comments*, and they're often included to make code more readable to human beings. The computer ignores them.) Once the code compiles, select Upload under the same menu. When you get the "Done uploading" message, you should see the red LED start blinking on and off, just like you programmed it to.

ARDUINO

SKETCH

VERIFY/COMPILE

STOP

ADD FILE . . .

IMPORT LIBRARY

SHOW SKETCH FOLDER

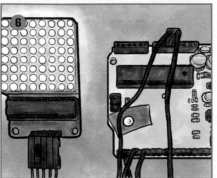

STEP 5 Disconnect. The really cool thing about what you just did is that you no longer need your computer to make it work; the program now lives in the microcontroller chip on the Arduino board and will run whenever the board has power. If you want to prove this to yourself, unplug the USB cord from the Arduino, then plug in the battery pack (with fresh batteries inside, obviously). The LED starts blinking again! To change how fast it blinks, all you have to do is change the numbers inside the parentheses in the "delay" functions of your program, reconnect the USB cable, and upload the new code to the board.

STEP 6 Blinking an LED is pretty tame, but it's a start. Traditionally in the computer world, when you're learning a new programming language, the first code you write consists of only one or two lines that output "Hello world!" onto the screen. Your Arduino doesn't have a screen, but you could easily write a program that flashes out the letters HELLO WORLD in Morse code with the LED. Try it out as your next challenge.

078 DON'T REINVENT THE WHEEL

Learning to code can be intimidating, so remember: It's not just you against the machine. If you're working with a manufacturer's development board, instructional materials and sample code will likely be provided, plus email or web contact points where you can send questions. If you're using an open-source product like Arduino, there's almost certainly an established online community where freely sharing and teaching are established norms. So when you're tackling a new programming problem, realize you almost never have to start completely from scratch. Even if somebody else hasn't shared code that does exactly what you want, you can find bits here and there that can be adapted to work for your purpose. Just be careful: While you can get away with pretty much anything during prototyping, when it's time to go to market, you'll have to thoroughly review your code to make sure you're not selling any pieces of code you're not allowed to. For this reason, it's a good idea to keep track of licenses as you go along, and always make sure you're in the clear *before* you copy-and-paste.

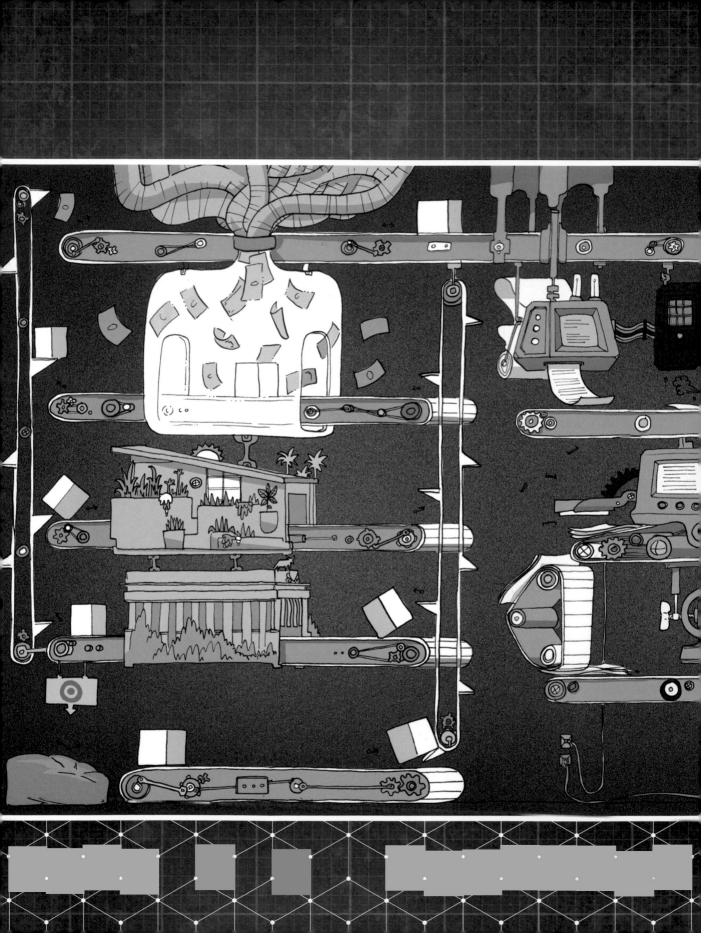

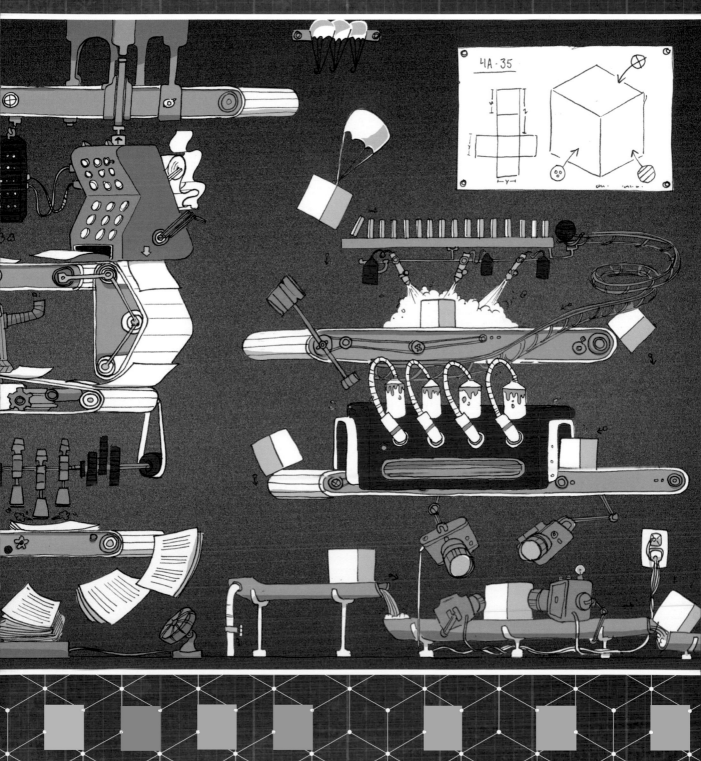

FINDING FUNDING

You've likely heard that there's no such thing as a free lunch. And for inventors, that meal can get prohibitively expensive—fast. Few can afford to self-finance the manufacturing, distribution, and marketing of their products. So if you're going to take yours beyond the working prototype, you're going to need to pass around your hat. Here's how.

:::

079 JOIN THE CROWDFUNDING CRAZE

If you need a million bucks, you can get it all from one investor, or you can get a dollar apiece from a million *microinvestors*. Of course, with so many investors, you'll spend a lot more time keeping people happy (to paraphrase Notorious B.I.G.: "Mo' backers, mo' problems.") Crowdfunding existed before the Internet, but came into its own with the launch of Kickstarter, a popular general-purpose crowdfunding site, in 2009. An online platform makes it easy to reach lots of potential microinvestors by sharing links on social media, and it automates the process of collecting their payments; in return, the site keeps a fraction of the money. Although the company has experienced growing pains, there's a lot of dough to be made here, and competition among platforms has become intense. If you're interested find the site that seems like the best fit for your idea (see #088). When in doubt, fall back on one of the big three: Kickstarter, Indiegogo, and GoFundMe.

080 TAKE IT TO THE BANK

If your needs are modest, the old-school approach—paying a visit to your local bank and applying for a small-business loan—can still work. In general, community banks and credit unions try to design their policies to benefit small-business owners, but you should expect banks to exercise more caution than other potential investors. You'll need strong collateral (which means you'll likely have to put up your house, savings, or other assets to guarantee the loan), plus good credentials and a strong personal credit history. You'll also need to be willing to wait as long as six months to get your money.

On the other hand, banks are likely to offer the best interest rates. Typical financing options from banks include commercial mortgages for buying or refinancing real estate, lines of credit for meeting

short-term operating expenses, and term loans for investing in fixed assets such as critical equipment and software licenses. Some government agencies, notably in the United States, will guarantee small-business loans to banks, meaning the government will pay back your loan even if you default. In rare cases, they'll do this even if you don't have strong collateral.

||||||||||||||||||||||||||||||||||||||||||||||||||||||||||||||||||||||||||||||||||||||||||||||||||||

081 CHECK OUT THE TOP KICKSTARTERS

Some crowdfunding campaigns have really hit the mark—and the jackpot. Kickstarters can serve as models for your own fundraising efforts, but keep in mind that some flopped in production.

US$10 MILLION

PEBBLE This relatively cheap e-paper watch with a long-lasting battery was the first to sync with phones, beating the usual tech giants to the smartwatch scene.

US$13 MILLION

COOLEST COOLER With a built-in boom box and blender, the Coolest Cooler went viral, but delays in production infuriated many of its 62,000 backers.

US$9 MILLION

BAUBAX TRAVEL JACKET The utilitarian hoodie that has it all—inflatable neck pillow, sleep mask, gloves, and nine pockets—was the top clothing Kickstarter.

US$8.5 MILLION

OUYA This Android-based gaming console for TV disrupted the industry with its low price tag and open invite to developers to make their own games.

US$6 MILLION

PONOMUSIC This portable music player developed by Neil Young exceeded its goal with rewards like devices signed by famous musicians and VIP listening parties.

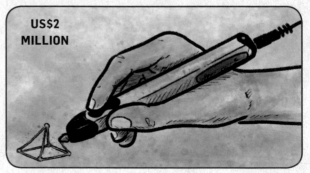

US$2 MILLION

3DOODLER A pledge of US$50 was all it took for backers to receive their own 3D-printer-in-a-pen. The target of US$30,000 was reached in a matter of hours.

082 GET AN ANGEL ON YOUR SHOULDER

In the finance world, an *angel* is a well-off person who invests her own money to help start your company (expecting that you'll succeed and she'll get that money back—and then some). If you have a rich uncle, he may be able to play the angel role, or you may impress a successful stranger by networking at a trade show or an industry conference. While a relative might offer you a better deal, the personal consequences could be harder to manage if your business flops. Regardless, if friends and family can't help you out, look around for an angel investor network in your hometown or online. The Angel Capital Association website is a good place to start.

083 PLAY THE VENTURE CAPITAL GAME

Unlike angels, *venture capital* (VC) *firms* aren't individuals—they're companies that represent pooled resources from many investors who may themselves be companies or people. The firms manage and (hopefully) grow that pool of capital by investing in startups they think will succeed. Many VC firms are connected to an *incubator*, a company that helps new businesses get off the ground by providing consulting, training, and sometimes physical resources such as an office or lab.

084 CALL UNCLE SAM

Or John Bull. Or Mother India. Or whatever dubiously familiar anthropomorphic form your particular nation-state happens to manifest. Many government agencies around the world make funds available to citizen entrepreneurs trying to get a new company off the ground. Just what options are available will depend on where you live and what kind of company you're trying to start, with generally more favorable odds if your answers are "a wealthy, industrialized country" and "defense-related," respectively. Start with your nation's small-business development agency and go from there.

085 DECIPHER FINANCIAL STATEMENTS

To grow a business, you've got to talk numbers. (Which is also a handy skill for breaking up parties that start to get too interesting.) Practically speaking, you should be familiar with three standard documents that record a company's activities and overall health.

BALANCE SHEET This document gives a snapshot of your financial position at a moment in time according to the accounting equation: Net worth = Assets   Liabilities. *Net worth*, aka *equity*, is the value of the company to its owners at that instant. *Assets* are the combined value of the stuff you own (like inventory, equipment, and patents), while *liabilities* are the combined value of everything you owe (like wages, taxes, and pensions).

INCOME STATEMENT Look to this document for a detailed breakdown of your income, expenses, and profits or losses over a period of time. The simplest type lists income on the top line and subtracts expenses to arrive at the bottom line. It tells whether a company made or lost money (and how much) but it doesn't indicate overall worth. A company can show positive income statements for a long time but still have negative equity (which means it hasn't made any money for its owners).

CASH FLOW STATEMENT This report explains how you paid your bills during a certain time period, and it can be used to estimate how you'll do so in the future. Even if your company is profitable and has a positive net worth, it can still get in trouble if it doesn't keep enough cash around to pay debts. If you spend all your money on equipment and can't pay your employees, your business is in trouble, no matter how it looks otherwise.

086 START YOUR BUSINESS PLAN

No matter how clever your idea, no savvy investor is going to fund you based on concept alone. (No, not even after she sees how good-looking you are.) She'll need proof that you can run a business and will want to see it in the language of business, which is accounting—aka, numbers. That means developing a formal document called a *business plan*, which charts the course of your company for its first five years and explains how you intend to generate revenue. This takes some doing, so get started early to avoid last-minute stress.

EXECUTIVE SUMMARY This part comes first but should be written last. In a single paragraph, briefly sum up the rest of the document. Write to persuade someone who reads a lot of business plans to dive further into yours. Quickly address your experience, background, and reasons for starting a company, plus any accolades or notable achievements. Make sure to specify what you need (i.e., your dollar amount) here too.

COMPANY DESCRIPTION What, exactly, is your business? Briefly summarize who buys your product and why as well as advantages you have over your competition.

MARKET ANALYSIS Explain how your industry has done in the past, how it's doing now, and how it's expected to do. Survey the customer groups in your industry, then narrow down to your target market: Who are these people, where are they located, and what do they need? How many of them exist and how much do they spend? How much of that market do you expect to capture? Then detail how you'll set your product's price.

ORGANIZATION & MANAGEMENT Write a list of the company employees and their responsibilities. What are their backgrounds and why are they qualified? If you've got more than eight staff members, include an organizational chart. Define

the legal structure of your company and identify its owners. If you have a board of directors, who's on it and why?

PRODUCT SPECIFICS Make your product's benefits clear. How does it compare to others on the market? Have you secured any intellectual property, like copyrights or patents? (See #096–100.) If not, are you trying? Demonstrate what stage of development you're in and how much more research and development you plan to do.

SALES & MARKETING Explain how you'll break into an established market, if there is one for your product. Detail how your salesforce will find prospects and convert them to customers. How will you distribute your product to those customers? How will you grow your company?

FUNDING NEEDS Be as detailed as possible about how much money you need and how you plan to spend it. Remember that you need to plan ahead five full years.

FINANCIAL PROJECTIONS Here's where you really get down to brass tacks. Include forecasted balance sheets, income statements, cash flow statements, and budgets for each month of your first year, each quarter of the second year, and each full year beyond that. And don't just hit them over the head with a bunch of spreadsheets. Give written summaries and analyses of the hard numbers, including charts and plots, so the reader understands—and values—your data.

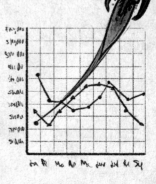

087 AVOID CRASHING AND BURNING LIKE THE ZANO

In November of 2014, a Welsh company named Torquing Group Ltd., launched a Kickstarter seeking £125,000 to fund its first run of Zano, a palm-size quadcopter drone. Zano had an impressively long feature list—gesture-based smartphone control, obstacle avoidance, automatic selfie videography—and an impressively low price: Just £170 would get you one, and you were to have it, Torquing said, by the following June. The Internet sat up and took notice, investing more than £2,335,000, a record-breaking sum for a European Kickstarter. And more than 20 times its goal.

The June deadline came and went, however, and no products had been shipped. Then, in early November of 2015, technical lead Ivan Reedman suddenly resigned. Torquing declared bankruptcy a week later, and the principals dropped off the web under a firestorm of indignant rage from disappointed backers. Only 600 Zanos were delivered, and those were barely able to get off the ground. Most of the 12,000 backers received neither drones nor refunds—and probably never will.

For Kickstarter, the Zano episode was a serious PR blow; The company had singled Zano out as a staff pick. So Kickstarter sent journalist Mark Harris to pore over Torquing's finances and speak with Reedman and other former employees. Expecting fraud, Harris found a different story: Zano's creators had been trying all along to live up to their promises. They just couldn't do it. Harris paints Reedman as a well-meaning visionary with a chronic tendency to overpromise and underdeliver. And he was the only one who understood how Zano was supposed to work. "The truth," Harris writes, "seems to be that almost everyone at Torquing was out of their depth." The takeaway here? Don't let your enthusiasm or ego drive the schedule, and surround yourself with people who can help and hold you accountable.

088 PICK A CROWDFUNDING PLATFORM

Online crowdfunding is a powerful tool, but it's not a one-size-fits-all solution. If you're trying to break into an established market, especially one that's very competitive (like consumer electronics), even the relatively fast cash-in-hand that you can win from crowdfunding may not come fast enough. In such markets, crowdfunding can still be a valuable tool for promoting your ready-to-launch product and scoring presales, but you shouldn't depend on it to get off the ground. Instead, you should raise private seed capital from family or angel investors (see #082), get your manufacturing pipeline set up, and then use crowdfunding to hype the launch, penetrate the market, and pay for your first production run.

The first important decision is going to be picking the right host, as not all crowdfunding sites are alike. Some are designed for charitable giving, whereby

the only reward is a sense of having helped a good cause, while others are based on tangible rewards: "In exchange for your investment, we'll send you a special-edition version of our product (or a sticker if you can only give a buck or two)." Some offer traditional equity investment, whereby the reward is actual shares of stock the company. Others offer what are called *debt securities*, where the operator makes a pledge to pay back each investment at a certain interest rate by a certain time. Some sites even beef up their usefulness with more services—like manufacturing help and order fulfillment—but ask for correspondingly higher cuts of your earnings. Some make you set a goal and won't deliver any money unless you hit it; others will let you keep whatever you raise. So do your research carefully before committing.

089 RUN A WINNING CROWDFUNDING CAMPAIGN

A word to the wise here: Crowdfunding takes a lot of work to, well, work in your favor. Here are the hallmarks of a successful campaign.

SOLID BUDGET How much should you ask for? First, figure out your *unit cost*, which is what you pay to make just one product, accounting for factors like shipping and overhead. Then determine the size of your initial run. You may have heard about *economies of scale*, which is a fancy term for the common-sense notion that unit cost goes down if you buy more at once. You'll have to crunch the numbers, but a good rule of thumb is to plan for 10,000 units. Once you have firm quotes from reliable partners, tack on a generous margin to guard against unlucky accidents. Common wisdom is to ask for no less than 115 percent of your projected costs plus the cost of using the site. Make your budget prominently visible and easy to understand on a line-by-line basis somewhere on your project's profile.

FAN BASE Crowdfunding campaigns usually grow out of the creator's personal social network. When your campaign goes live, the first thing you'll do is ask your contacts to share your project, so it makes sense to beef up your own network on major social media platforms prior to launch. It makes even more sense to join communities where your potential customers hang out, identify *influencers* (users with a lot of status among your customer demographic), and come up with strategies for catching their attention. It's okay to go back and ask your family and friends to share or contribute as the campaign unfolds, but if you spam people every day with requests for money, you'll quickly lose their interest. (And possibly their friendship.) Everyone wants their campaign to go viral, but it's that core community of fans who will make or break you, so speak to their needs first.

COMPELLING STORY Your goal in building a campaign website should be to tell an engaging narrative about yourself, your invention, and your adventure bringing it to market. Besides the just-the-facts stuff (who, what, when, where, why, how, how much), you need to connect with people emotionally and get them excited about your product. The best way to do this is to make you—yes, you—the spokesperson of your campaign.

CONSISTENT PRESENCE If people comment on a website, via email, or in a conversation on social media, don't leave them hanging. Respond quickly, informatively, and positively, even if their tone is critical or downright rude. And once you've got people's money, don't forget about them! Thank everyone who backed you in the most personal way practical, then stay in touch by publishing regular updates about your progress. There's no hard-and-fast rule about how often to post updates, but once a week is a sensible target. Think of these people as partners—if you encounter problems, let them know, along with what you're doing to solve those issues. You should also monitor your campaign's page and trending status on social media. If things aren't taking off, you need to actively promote the project without making a nuisance out of yourself.

TIMELY DELIVERY Unless you are absolutely positive about your production timeline, it's usually a mistake to promise results by a specific date. Instead, use language like, "We expect to start shipping our product in the third quarter of the year." This gives you a soft goal (ship one unit), and a wide margin of error (three months) in which to hit it. Delays can make for bad publicity, so pad out your schedule as much as possible.

090 MAKE A KILLER KICKSTARTER VIDEO

To spread the word about your campaign online, you'll have to produce supporting text, images, and video. Of these, video is most important, because it's the most popular and easy to share. Making a good video usually takes more time and energy than creating good still photos or supporting text, but it pays off. It also takes some tools: an HD video camera, a tripod, a wireless lapel mic, some sort of lighting gear (however rudimentary), a personal computer, and video-editing software.

The trick to making good videos is to watch lots of them. Look up some successful campaigns on your platform, watch their videos, and ask yourself why you like some more than others. Now make a video that you, yourself, enjoy watching.

STEP 1 Start by making a list of everything you want to say. First, the facts: who, what, where, why, when, how, how much. Then add details about your story. How did this idea come to you? Why is the product special? Now turn this list into a script for dialogue (on camera) and voice-over (off camera). Edit this script hard, leaving only the most important bits, expressed in short and clear words. Scenes with actors are tough, so use as much voice-over as possible. Once the script is tight, imagine what viewers will see at each moment. This is your shot list.

STEP 2 Lighting is the most important factor in video, much more so than the quality of your equipment. Aim to illuminate your scenes so no harsh shadows fall on your subjects. The easiest solution is to shoot outdoors on an overcast day, especially during the *magic hour*: the periods before sunset and after sunrise when the light takes on a golden hue. If you must shoot inside, black out natural light and use clamp lamps with full-spectrum bulbs to illuminate your scene, hanging up white sheets to reflect or diffuse light and to cancel out shadows.

STEP 3 Record your on-camera dialogue with a good wireless lapel mic. The actors—either yourself or others—should look directly at the camera lens and talk to it at a natural pace, as if it were a good friend. Be patient with flubbed lines (yours included) and be prepared to reshoot if you're in a busy place with lots of background noise.

STEP 4 Now shoot the scenes that don't require live audio. This footage is often called *B-roll*, and it can be of your office or factory, early prototypes, people happily using your invention, whatever. Later, instead of reshooting an imperfect scene, you might be able to edit around it by cutting away to B-roll. (It's also okay to use still images instead of moving pictures, so long as they're interesting.)

STEP 5 Next, find a nice, quiet place to record the voice-over (the lines that will be layered over the B-roll). Here, too, background noise may be a problem. If you don't have access to a sound booth, turn off any central-air systems, unplug noisy appliances, and retreat to a walk-in closet full of clothes—all that heavy hanging fabric does a great job of absorbing extraneous noise.

STEP 6 Edit. And edit some more. This will be easiest if you invest in some cheap but decent nonlinear (aka nondestructive) editing software. You can buy a month of access to Adobe Premiere Pro for about the price of an average dinner for two, or you can check out one of the "freemium" programs like Lightworks. Once you dive in, you may have to go back, reshoot, or rerecord here and there—this is natural. Avoid special effects like fancy title cards or wipes until you know what you're doing. Make it short and sweet: People should know what you're about within the first ten seconds, and the whole thing should be no longer than a minute and a half.

STEP 7 Once you've got a video file ready to go live, don't slack off on the details of getting it online. Video-hosting websites have scores of options, and they can all have important consequences for how easy your video is to find, watch, and share. Don't skip providing high-quality subtitles that are synced with the video—besides the hearing-impaired, viewers may want to watch your video with the sound off. Other factors: Are you going to allow comments? If so, are you prepared to respond to them? What about embedding? Will people be able to play your video from any webpage, only from a particular page, or only from the video-sharing site itself? Make sure you choose the best settings before going live.

091 JUST ADD CATS

There's a concept in marketing that calls for offering "candy with the medicine," meaning your message is best received when accompanied by a healthy dose of entertainment. What makes a video go viral? While there's no foolproof formula, these factors help: a memorable story, a catchy title, a solid sense of humor, a little bit of wackiness (think fun, not disturbing), appealing to universal experience, and yes, a cute creature or two—be they cats, dogs, bunnies, or babies.

092

MEET HELEN GREINER

CEO OF CYPHY WORKS, COFOUNDER OF IROBOT

Helen Greiner was 11 when she saw *Star Wars: A New Hope* on the big screen. She found herself captivated not by the smooth moves of Han Solo or the strength of Princess Leia but by R2-D2. "I was enthralled," Greiner says. "He had a personality [and] an agenda, and [he] was a main character. He was more than just a machine. For me, he was a muse."

When her father brought home a TRS-80 (aka, the "Trash-80"), Greiner—realizing it could be the path to creating robots like R2-D2—co-opted it as her own, teaching herself to write code. In high school, she was the best computer programmer in class, but when it came time to go to college, she was disappointed to find that MIT didn't have a specialty program in robotics. She settled for an engineering degree followed by a master's degree in computer science.

In 1990, Greiner was finally ready to realize her dream of building robots herself. She cofounded iRobot, a company that eventually sold millions of home robots as well as thousands of machines for the U.S. military. In the beginning, however, that success seemed far from guaranteed. The company was small, and robots were not popular in those days, instead seeming the stuff of science fiction. Greiner spent her time creating robots for the first five years after cofounding the company, until she realized that "the other business stuff wasn't getting done." She and her cofounders had started iRobot to pursue "a vision of building robots," but they hadn't formulated any sort of practical plan for doing that. "That's the wrong approach to starting a business," she says. "You should have a vision, but you should also have a business plan behind you."

Cash strapped, they managed to raise some venture capital in 1998 to supplement the

continued on next page

money they were making with their early products. A couple years later, iRobot came up with an idea for a robotic vacuum cleaner called the Roomba. While other companies had taken stabs at robotic vacuums, the iRobot design was new and, in Greiner's opinion, superior.

But she and her colleagues had no idea if buyers would agree. The Internet was still young in 2000, and platforms like Kickstarter—which are used today to test public interest in new ideas—certainly didn't exist. So Greiner hired a professional firm to recruit a focus group mostly consisting of women, the Roomba's target audience. "At first they were imagining the Terminator pushing around a vacuum," Greiner says. "But when they actually saw the Roomba, they were like, 'Oh yeah, I could see that in my house; it won't take up the whole closet.'" The women also approved of the Roomba's low price tag. That was all the encouragement iRobot needed. Since the Roomba's release in 2002, more than 10 million units have sold.

In 2008, Greiner founded CyPhy Works, a startup that specializes in drones for personal, military, and commercial use. Her innovations there and at iRobot have earned her numerous awards, including one of America's Best Leaders, from the Kennedy School at Harvard University, and the Pioneer Award, from the Association for Unmanned Vehicle Systems International. Though it has been decades since she first saw R2-D2 on the big screen, her enthusiasm for robots shows no signs of waning. "I'm still so excited about robots," she says. "And now it seems that the rest of the world has finally caught up. It's a really, really exciting time for the field."

Q: What's your earliest memory of tinkering?

A: I started digitally with a TRS-80 computer when I was 11. At the time you had to program it in BASIC and make graphics with ASCII characters and text semigraphics. I was an early digital native. Before this, I remember trying to build a robot out of an old typewriter. I didn't get very far. That was third grade.

Q: What's your favorite material?

A: Unobtainium because it has infinite strength and no weight, and ABS because it can be injection-molded into a vacuum or a kayak. That's versatility.

Q: What's it like to see your invention in the wild? Are you ever surprised by its uses?

A: Customers have always said, "I want a voice-activated Roomba." Then we ask how much they would pay for it, and usually the answer is very low, like US$10. Now customers with Amazon Echo or Google Home should be able to get this functionality for free.

Q: What's your dream invention—if time, money, or the laws of physics didn't apply?

A: A time machine to go look at future inventions. The past is overrated! The best time is now and, I hope, the future.

Q: Are there any other inventors or makers who particularly inspire you?

A: Limor Fried, aka Ladyada—she helped start the maker movement.

Q: What happens when you get stuck working on an invention? How do you hit refresh?

A: Talk to people. Seek out-of-the-box thinking. Good ideas can come from anywhere.

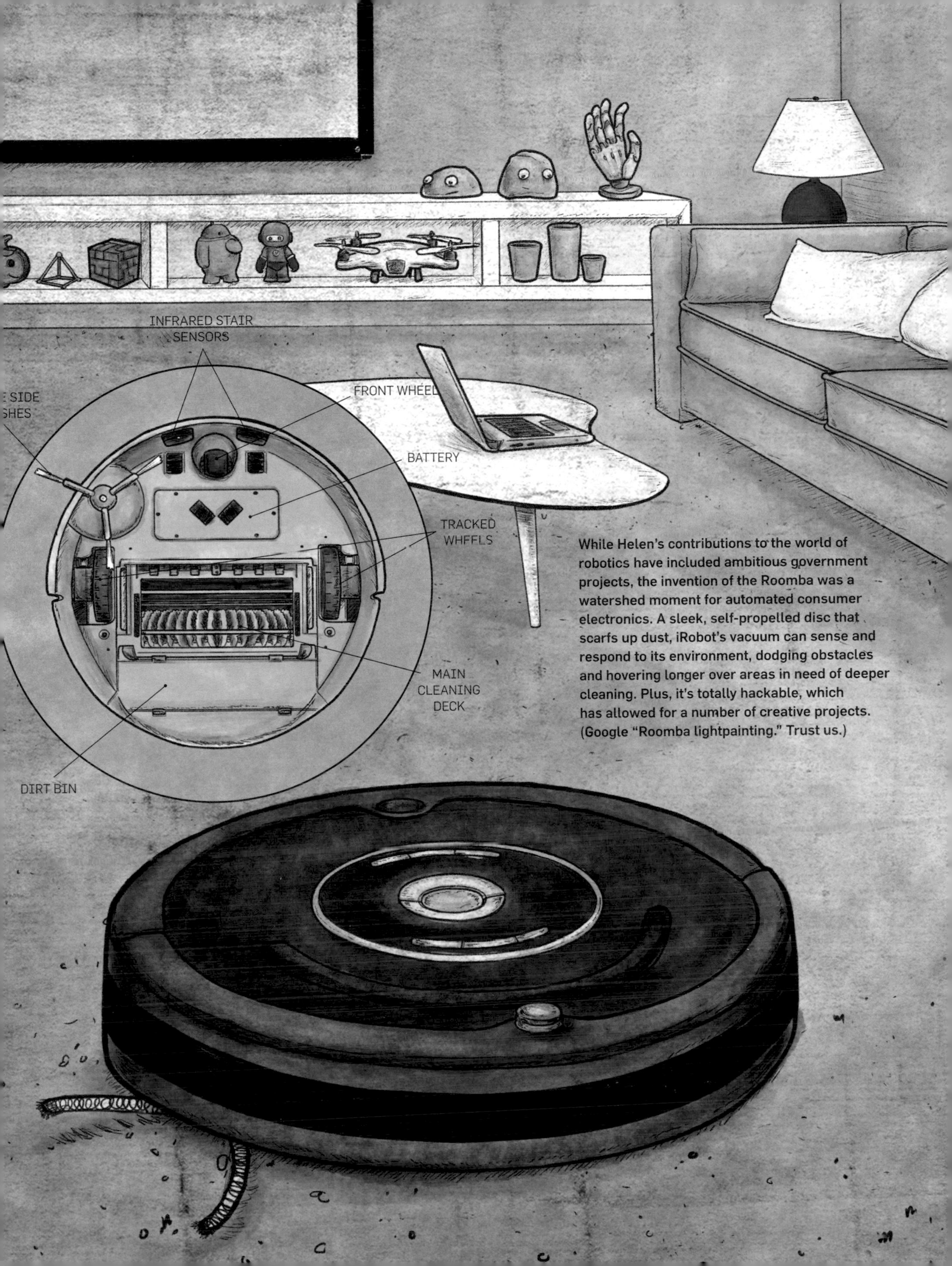

INFRARED STAIR
SENSORS

E SIDE
SHES

FRONT WHEEL

BATTERY

TRACKED
WHEELS

MAIN
CLEANING
DECK

DIRT BIN

While Helen's contributions to the world of robotics have included ambitious government projects, the invention of the Roomba was a watershed moment for automated consumer electronics. A sleek, self-propelled disc that scarfs up dust, iRobot's vacuum can sense and respond to its environment, dodging obstacles and hovering longer over areas in need of deeper cleaning. Plus, it's totally hackable, which has allowed for a number of creative projects. (Google "Roomba lightpainting." Trust us.)

093 GET INDUSTRY ADVICE ON GETTING FUNDED

Securing funds for your project is an intimidating undertaking, no doubt. Here's some guidance from folks who've been on both sides of the table.

On the selection process for funding startups: "Usually what we're looking for is a team that has some engineering skill, maybe some design and marketing skill. We like to see that there is a functional product that's already been built; it doesn't have to have a lot of customers or usage but at least demonstrates that they can build a product. And then we want to understand how they make money, at least have a good idea about how it could make money. Then we usually make decisions somewhat on gut, and then 6 to 12 months later we find out whether we're right or wrong [. . .]. And then if we're right, then we keep investing in the company and hopefully we find a few that work."

— Dave McClure, founder of 500 Startups

"STAY SELF-FUNDED AS LONG AS POSSIBLE."
—Garrett Camp, founder of Uber and StumbleUpon

"WHEN YOU FIRST MEET AN INVESTOR, YOU'VE GOT TO BE ABLE TO SAY IN ONE COMPELLING SENTENCE WHAT YOUR PRODUCT DOES."
— Ron Conway, angel investor and early investor in Google, Ask Jeeves, and PayPal

"YOU'RE ALMOST ALWAYS BETTER OFF MAKING YOUR BUSINESS BETTER THAN YOUR PITCH BETTER."
— Marc Andreessen, entrepreneur and cofounder of Netscape

"Ask any venture capitalist, and they will tell you that they consider the experience and completeness of the founding team to be a more important factor in their investment decision than the technology that is being built."
– Vivek Wadhwa, entrepreneur and *Washington Post* tech columnist

"You need to begin to network with angels and VCs while you are still ideating. It is easier to ask someone you know for funding than a stranger. Build your financial network by attending as many industry functions and reaching out for advice from experts online." **– Jay Samit, digital media innovator**

"Everything about my journey to get Spanx off the ground entailed me having to be a salesperson— from going to the hosiery mills to get a prototype made to calling Saks Fifth Avenue and Neiman Marcus. I had to position myself to get five minutes in the door with buyers." *– Sara Blakely, inventor of Spanx*

"CHASE THE VISION, NOT THE MONEY. THE MONEY WILL END UP FOLLOWING YOU." *–Tony Hsieh, Zappos CEO*

"Don't just take any deal. Be selective and treat each deal as the last one you will ever have. Take nothing for granted. And keep taking chances; that's how things get done." *– Alan Amron, inventor of the Post-It Note*

"Launch your product or service before you have funding. See how people respond to it before you have a PowerPoint and business plan—have something people can use, and go from there." – Chad Hurley, cofounder of YouTube

MEET PETER HOMER
INVENTOR OF THE ASTRONAUT GLOVE

Peter Homer never thought of himself as an inventor, but when he stumbled across an online ad for NASA's 2007 Challenge for Improved Astronaut Gloves, his interest was piqued. NASA sought a solution to a perennial problem for astronauts: The force required to move their fingers and wrists within pressurized gloves took a toll, often causing blisters, abrasions, damaged fingernails, or cramps. As NASA explained, "New technologies would reduce discomfort and make the astronauts' jobs easier and safer."

Homer realized that his background made him uniquely qualified to take a stab at solving the glove conundrum. His degree was in mechanical engineering, and his first job out of college was at an aerospace company working on satellites. Later, he transitioned from hardware to software, adding computer programming to his skill set. In between those gigs, he had also taken a year-long position selling sails, so he was familiar with various fabrics. To boot, he already knew how to sew.

"The NASA glove challenge looked like something I could do in my garage," Homer says. "I didn't have a whole lot of equipment or money, but I figured I'd give it a shot."

Homer joined other competitors at an introductory meeting NASA held in Connecticut. There, he had the chance to try a pressurized glove situated inside a contraption called a glove box. The box simulated the conditions in space by exerting 4½ pounds of pressure per square inch (0.3 kg/sq cm) on the glove inside. Homer was surprised to find that, under pressure, the soft material turned rigid, giving the glove the feel of a knight's metal armor, and the fingers resisted movement as though they were spring-loaded. "I was given three minutes in the glove box, and after that I went off to a corner and wrote for about 15 minutes about my impressions of what I'd just felt and witnessed," Homer said. "I was really surprised by what I felt—it was totally unexpected."

continued on next page

Back home, he launched into a hands-on design approach, creating quick-and-dirty prototypes for testing. He ordered bolts of lightweight polyester and raw metal materials from eBay, and he used a bell jar combined with pieces from a blood pressure pump to rig up his own version of the glove box. "My first glove was a disaster," he says. "It didn't even look like a hand when pressurized."

About six weeks before the deadline, he had an epiphany and realized that all he really needed to do was build fingers that performed better than the agency's existing model. Homer's dining room table became cluttered with 25 or 30 fingers, each slightly better than the one before. Finally, two weeks before the deadline, he landed on an acceptable design. He spent the remaining time building out the full glove, along with a requisite duplicate for the left hand. He finished two days before the deadline.

Despite the rush job, Homer's glove won, netting him US$200,000. The news came as a surprise to him but also a relief; he'd recently been laid off, and he had been working on the glove instead of looking for a new position. Now he planned to get on with the job hunt, but he received a call from Rick Tumlinson, cofounder of the Space Frontier Foundation, a nonprofit organization dedicated to turning humans into a multiplanetary species. Tumlinson wanted to hire Homer to work as a consultant designing a full space suit. Homer agreed.

Homer also floated the idea of licensing or selling his glove to a number of companies, but all thought the market was too limited to invest in space suit technologies themselves. Instead, they preferred contracting with Homer. It seemed there was enough work, so Homer founded a company, Flagsuit (an acronym for "fits like a glove"). Customers include several U.S. space suit companies as well as NASA. Designs have improved over the years; in 2009, Homer again won NASA's space glove competition, this time by a much wider margin. "When I won the first competition, my glove received about a 3 percent better score than the NASA glove," he says. "But at the second competition against the same NASA glove, mine was more than twice as flexible."

So far, the company is doing well, though Homer prefers the creative work to the drudgery of accounting that inevitably goes hand in gloved hand with running a business. "The biggest challenge is making sure I can put food on the table for my family while still getting to play in my sandbox," he says.

Q: What's your earliest memory of tinkering?

A: I remember, when I was nine or ten years old, taking apart a toy train and using the guts to make a set of motorized windshield wipers that fit on top of my eyeglasses.

Q: What was your greatest challenge in making your invention a reality?

A: My greatest challenge is that I enjoy the creative process more than company building, fundraising, or marketing. Even though I know these things need to be done, I find myself gravitating back to inventing and building.

Q: Are there any other inventors or makers who particularly inspire you?

A: I think Nikola Tesla is cool because he focused on solving real-world problems, not just coming up with a product around which to build a company.

Q: What's your dream invention—if time, money, and the laws of physics didn't apply?

A: I'm a huge advocate for human settlement of space—not just the planets and moons but also self-contained city-ships in orbit around the Sun. I'm interested in working on anything that helps this become a reality sooner.

Q: Do you have any words of wisdom for aspiring inventors?

A: Break it down to the essential problem that needs to be solved. Don't overthink it—get your hands dirty early! I learn so much more—and so much more quickly—by trying to make things and failing than by just analyzing.

Q: If you could go back through your invention process, what would you do differently?

A: Start building a lot sooner!

Innovating for life on Earth is one thing; making products for life outside our atmosphere is another. Here are some of the features that helped Homer's space-suit glove outshine the competition.

A curved metal hoop around the base of the thumb gives it the ability to move in all directions.

Bent metal rods cross the knuckles to form a single metacarpophalangeal joint, which helps with flexibility and keeps the glove from ballooning out across the wearer's palm.

Crossed reinforcements on each side of each finger joint create a hinge, which simultaneously allows rotation and prevents separation of the joined finger elements.

Due to its design and fabric patterning, Homer's astronaut glove is rigid enough to be the same shape, even when there's no hand inside or when it's lying on its side.

SWIMMING WITH SHARKS

Lawyers! Who needs 'em? Well, like it or not, you do, if you're going to develop and protect an invention, then start and run a company to produce said invention. A good attorney will help you navigate the ins and outs of protecting your shiny new concept, including officially getting yourself incorporated and filing for that crucial patent.

:::

095 LISTEN TO THE GRAPEVINE

When you're looking for an attorney, think twice before calling a number off a billboard or clicking an online ad. The best lawyers rarely advertise, for the simple reason that they don't have to: Word gets around, and all the clients they're ever likely to need come knocking. The best way to find these people is to work your connections—talk to friends, family, coworkers, or successful entrepreneurs in your area, and get some recommendations based on firsthand experience.

When you get a lawyer on the phone, tell him where you got his name, ask if he's taking new clients, and briefly describe what you're trying to do. He'll take it from there. Remember: You're interviewing him, not the other way around. You want someone with experience in business and intellectual property law; criminal lawyers, estate lawyers, and family lawyers are right out.

Trust your guts and politely pass on anyone who rubs you the wrong way. Trust your head too, and don't get snowed by somebody who can't or won't answer all your questions in clear and easy-to-understand words. Even if the first lawyer you talk to knocks your socks off, do a bit of comparison shopping before deciding.

096 PROTECT YOUR BRAND

Intellectual property, or IP, refers to a type of asset that has value even though it has no physical form. Case in point: 30 years ago, *google* was a word only a mathematician or Scrabble champion had ever heard. Now, almost everyone recognizes that name as one of the biggest companies on Earth. And that recognition is a valuable asset to Google, which has a correspondingly strong interest in preventing anyone from misrepresenting themselves as *the* Google. To prevent this kind of damage, laws all over the world protect the right to exclusively use certain symbols as trade- and service marks. There are three basic types.

TRADEMARKS These protect the words, phrases, symbols, designs, and combinations thereof that identify goods as coming from one source rather than another. The symbol ™ indicates an unregistered trademark, which you can legally establish just by regularly using the mark in your business.

SERVICE MARKS You can use this mark to identify services instead of goods. The symbol ℠ indicates an unregistered service mark. Otherwise, service marks are identical to trademarks in every way.

REGISTERED MARKS These protections cover the same kinds of IP as the unregistered types. To use the registered mark symbol ®, you must apply for and be granted rights to that mark with a national government. This costs money but provides wider and stronger legal protection than an unregistered mark.

097

PROTECT YOUR HARDWARE

When you develop a new and useful invention, you've created intellectual property. The knowledge of how to make and use your invention didn't exist before you figured it out, and that knowledge has value apart from any objects you build along the way. You should protect that knowledge so others can't profit from it unfairly.

If you're selling a physical product, it's usually easy for competitors to take it apart and reverse-engineer it to figure out your design. (In fact, we suggest you do just that in #064.) If those competitors could then turn around and sell your invention, people would be discouraged from inventing new things. So governments usually offer inventors special sets of rights, called *patents*, in exchange for full public disclosure of the technical details of their inventions.

Getting a patent is a months- or even years-long process in which the government makes sure the invention is new and useful (see #106–111 for a step-by-step description). If granted, it allows the holder to sue anyone who replicates her invention without permission. Once the patent term expires, the invention is fair game, but before it does, the inventor enjoys what amounts to a decades-long legal monopoly.

098

PROTECT YOUR SOFTWARE

Though patents for software have been granted in the United States and other parts of the world, the patentability of software inventions is nowhere near as clearly established as that of physical inventions. Generally, computer programs are protected under the third major type of intellectual property, which is *copyright*. Just as an author has a right to prevent others from selling unauthorized copies of her work, and a musician has a right to prevent others from selling unauthorized copies of her songs, a programmer can claim copyright to prevent the unauthorized distribution of her computer code. Your company will also own copyrights on, for example, any photographs, text, or videos you produce to promote or explain your products.

099

PROTECT YOUR POSTERIOR

Besides securing your intellectual property rights, you need a lawyer to help keep your company out of court. An attorney with a long history of winning lawsuits may seem impressive at first, but stop to think about how expensive all that court time was. The mark of a good attorney is that she can protect your rights effectively *without* ever having to set foot in court. While in the United States your company can literally be sued by anyone, for anything, no one is going to spend the time, money, and energy to pursue a case they don't stand a real chance of winning. Your attorney can tell you how to avoid taking actions or setting policies that are likely to expose you to the risk of losing a lawsuit in the first place, as well as how to respond in the event that someone actually sues or threatens to sue. Don't wait until you're panicking to find a good attorney; plan ahead and know who you're going to call if and when the worst happens.

HISTORY LESSON

100

DISCOVER THE FIRST PATENT LAWS

Generally, historians date the first modern-style patents to the glass-making industries of Renaissance Italy, 'round about 1450, and trace their development through France in the 1500s, England in the 1600s, and thence throughout the world by colonization.

U.S. Patent #1 was awarded on July 31, 1790, to one Samuel Hopkins of Philadelphia for a process used to make the chemical potassium carbonate. It's signed by George Washington. In April 2015, the modern Patent and Trademark Office (PTO) issued U.S. Patent #9,000,000 to a Florida man for a method of recycling rainwater that runs off a windshield as wiper fluid. We've come a long way.

101 START A COMPANY (IN DELAWARE)

While many modern companies operate around the world, corporations and other business entities are governed by laws that are relatively local. In the United States, corporations are chartered by the states, and Delaware has emerged as the favorite court system for businesses settling legal disputes. That's why more than half of U.S. public companies are incorporated there. Your local laws may well differ, but understanding how it's done in Delaware gives a good general overview.

STEP 1 Your business can't register in a place where it doesn't have a physical presence, so get an address in Delaware (or the area of your country that has the friendliest corporate laws on the books). When you're just starting, you can probably register with your own state using your own home, office, or other local address. If you want to incorporate in another state, you can usually pay a registered agent who works there to serve as your physical representative.

STEP 2 Figure out what kind of business you need. Are you trying to start a corporation, limited liability company, limited partnership, or something else? Choose wisely (see #102) because it can matter a lot as your business grows, especially at tax time or if you ever go to court.

STEP 3 Reserve the name. Think carefully because this will be hard to change later. You want a name that is evocative, memorable, easy to spell, easy to find online, and not too similar to an existing business name. Most states provide a web search tool that lets you check if a particular name is already reserved, and many will allow you to submit an application for reservation online too.

STEP 4 Fill out the form. Depending on what kind of business you're starting, this may be a Certificate of Incorporation, Certificate of Formation, Statement of Qualifications, or something else. The form will ask for basic information like the name of the business, its local address (or the address of its agent), the date, the name and address of the person filling out the form, and finally, that person's signature.

STEP 5 Pay the piper. Once your business is registered, you'll have to pay recurring taxes and fees on an annual and possibly quarterly basis. If you've formed a corporation, you'll almost certainly have to file an annual report, likely at the same time your taxes are due. In many places, it's possible to do all this yourself. But to avoid rookie mistakes that could be expensive later on, it's best to hire a lawyer, if you can afford it.

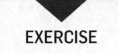

Welcome to
Incorporateburg
Land of the Cheap Business Charter

102 PICK THE BEST TYPE OF BUSINESS FOR YOU

While we often use words like *company* and *corporation* interchangeably, there are subtle but potentially crucial legal differences between the types of business entities recognized by governments around the world. (Read: Wake up! This could cost you money.)

SOLE PROPRIETORSHIP This type consists of a single person. He may file for a DBA (Doing Business As) name so that he doesn't have to use his real name, but otherwise there's no legal difference between the business and the person. The assets and profits are all his, as is the responsibility for any debts.

GENERAL PARTNERSHIP In this arrangement, two or more people share equal ownership and personal liability for debts and other legal obligations.

LIMITED PARTNERSHIP This situation consists of two or more people with at least one general and one limited partner. A general partner shares in the management and ownership and is personally liable for its debts and other legal obligations. A limited partner has limited liability, meaning she can't lose personal assets beyond what she invests, but she also doesn't have managerial authority.

LIMITED LIABILITY COMPANY The owners aren't personally liable for company debts but are personally responsible for paying taxes.

JOINT-STOCK COMPANY Here, ownership isn't shared equally but is apportioned among shareholders by investment in stock. Today, most joint-stock companies are also corporations, meaning the owners aren't personally liable for company debts and obligations, but that's not always the case.

CORPORATION Whether a sole proprietorship, a partnership, or a company, a corporation is any business that's legally recognized as a separate "person" apart from the people who own and run it. The corporation enters into contracts, pays taxes, and carries debts and duties apart from its owners.

103 PROTECT YOUR PEOPLE

If things go well, you'll eventually have to hire some help. Bringing on your first full-time employee is a bit like fording a river in the backcountry: If you know what you're doing and plan ahead, there's nothing to be afraid of—it'll be a fun adventure. But if you charge in without thinking, you can get in trouble fast. A business exists to make money, sure, but both from a moral and a legal point of view it also exists to provide a measure of security for the people who keep it running: owners and employees.

Though the laws vary around the world and it's crucial to study those in your area, the act of hiring people on salary usually involves entering into a binding legal relationship with them that carries certain rights and duties, on their parts and yours, that you need to understand before you sign. Even choosing one person over another can have legal consequences: Someone who can prove that he is qualified for a job that you gave to somebody else might, for example, be able to accuse you of discrimination or otherwise engaging in illegal hiring practices.

In the United States, when you make the decision to hire, you'll have to take steps to comply with a number of federal laws. You'll need to get an Employer Identification Number (EIN) from the IRS, set up a payroll system, arrange for federal income tax withholding, verify a hire's immigration status, make sure your workplace is safe to federal standards, post required notices, and pay unemployment taxes. At the state level, you'll need to register with your state's labor department, report the new hire to the agency that tracks child support, and buy worker's compensation insurance.

Before conducting any interviews, speak with a lawyer so you understand all the requirements of your area. Ditto in the unhappy scenario that you need to fire somebody; that's a delicate situation that can expose you to a lot of legal risk if handled badly.

104 DRILL FOR PROFITS BETTER THAN THE DRAKE OIL WELL

It's not news to anyone that a buck or two has been made in petroleum over the years. And though the use of oil that naturally seeps to the surface has been known since ancient times, extracting it in industrial quantities requires drilling. The modern petroleum industry dates to the introduction of drilling and is only about 150 years old. Today, oil and gas generate hundreds of billions of dollars in revenue each year, and it's no exaggeration to say that oil has greatly shaped the world we know—it creates fortunes, founds dynasties, and starts wars.

And all of that began on August 27, 1859, beside a creek outside Titusville, Pennsylvania, when a former railroad conductor named Edwin Drake literally struck it rich—69 feet (20 m) down in the ground. It had been hard going for the first 32 feet (10 m). The earth above the bedrock was soft, and pressure from groundwater kept collapsing the walls of the hole. To get around this problem, Drake ran a

cast-iron pipe down the borehole to hold up the walls. And though historians now know that he wasn't actually the first person to do this, he was certainly first in the United States, and the wealth and publicity his strike generated would've made patenting close to a sure bet.

But Drake, as historians politely observe, "lacked business acumen," which is a nice way of saying he made arguably the most expensive IP blunder in history. He had a chance to patent oil drilling, but didn't even try. Drake's historic strike still made him a small fortune, which he proceeded, in short order, to lose. In 1872, the state of Pennsylvania voted to award him US$1,500 a year in recognition for his historic contribution to an industry that made so much money for so many other people. Without this act of charity, he probably would have died in poverty. Don't make his mistake. Even if patenting seems silly or wasteful at first, the choice to skip it could haunt you.

105

MEET JULIO PALMAZ
INVENTOR OF THE STENT

Julio Palmaz, a newly minted vascular radiologist, knew he wanted to change medicine. In 1977, that drive took the native Argentinean to the States, where he began a residency at the University of California, Davis. "I essentially had an attitude of wanting to invent something," he says. "It was different from other inventors who just stumble on things. I was looking for opportunities."

Opportunity arrived at a lecture Palmaz attended, delivered by German radiologist Andreas Gruentzig. Several months earlier, Gruentzig had performed the world's first successful coronary balloon angioplasty. The procedure clears blocked arteries by advancing a catheter over a wire containing a balloon through blood vessels and then inflating the balloon at the site of the blockage. Doctors then remove the balloon and wire after the vessel is expanded. But the procedure was not perfect: As Gruentzig noted, blood vessels tend to experience elastic recoil, shrinking back to their original, blocked state after the balloon is removed.

Palmaz, however, believed he had a fix. "Right there, in that lecture hall, I came up with an idea to use some sort of mesh to reinforce the wall disrupted by the balloon, and then leave the mesh in place as an implant," he says. "That was 1978, and the idea stayed with me rather obsessively, I have to admit."

Indeed, it wasn't until 1990 that the first commercial stent was approved, though Palmaz began work on the project almost immediately after seeing Gruentzig's talk. He started in his garage, making calculations of the physics and geometry involved in placing a stent and creating prototypes out of cardboard. Eventually, the product got far enough along that he switched jobs, heading to the University of Texas Health and Science Center at San Antonio, a research

continued on next page

institution that would allow him to pursue the stent's creation during work hours as well as advance into the animal experimentation phase. "It required a fair amount of sacrifice for my family," Palmaz says. "We had to pick up everything and move across the country."

But even after he had a working prototype, the companies he approached declined to back the product. Palmaz, however, believed in his invention. "I knew that the device would do the job needed: to keep an artery open at 100 percent of its diameter," he says. "The more I worked with patients, the more I was convinced that this thing was needed."

Finally, the health-care company Johnson & Johnson came through, at first licensing the technology from Palmaz for US$10 million plus royalties and also investing in its development, and then later buying the patent.

Even then, though, Palmaz's struggles over the stent were not over. Several other companies soon began manufacturing stents, and Johnson & Johnson decided to pursue litigation against them. The problem came down to timing: Palmaz had waited until 1985 to apply for a patent, but by then two or three other people had already submitted similar applications for competing ideas. The legal battle went on for 12 years, but in the end Johnson & Johnson won. That victory was sealed only because Palmaz had written a report in 1978 describing the theoretical basis of the stent, and at the time U.S. patent law was based on the first to have an idea, not the first to patent it (which is no longer the case). The damages awarded to Johnson & Johnson at the end of the battle—US$3.6 billion—were among the largest in patent litigation history.

Palmaz is now retired from vascular radiology, but he continues to work on improving the stent, focusing especially on creating smaller devices for very fine vessels found in critical areas like the brain. He believes such microscopic stents could help prevent strokes, a medical problem he is passionate about helping to solve. "I always say the inventor has an obligation to keep working on his invention for life," Palmaz says. "Invention is never done."

Q+A

Q: What's your favorite tool?

A: TIG welder.

Q: What's your earliest memory of tinkering?

A: I remember being a young child fascinated with clockworks and taking them apart.

Q: What was your greatest challenge in making your invention a reality?

A: People's reluctance to consider placing a metallic implant in vessels. Before I even got a prototype, every expert I consulted considered the stent a terrible idea.

Q: What's your favorite material?

A: Today: high-strength composites. In 1978, stainless steel 316L.

Q: What do you do when you get stuck working on an invention? How do you hit refresh?

A: The best way to approach a project is to consider it a failure before you start. Nothing works until proven otherwise. You have to be humble and patient.

Q: What's your dream invention—if time, money, and the laws of physics didn't apply?

A: Totally automated nano manufacturing.

Q: Do you have any words of wisdom for aspiring inventors?

A: Learn to challenge the status quo as a habit. Invent for fun and not for fame or money. Those come as consequences and rarely as the result of pursuit. Learn to trust only yourself and your instinct. Learn to respect your intellect and your creative work. Be patient, humble, and focused. Stay light on your journey.

Every year, heart disease kills more than 17 million people worldwide. How much worse would those numbers be if it weren't for the stent, Julio Palmaz's treatment for clogged arteries? When a buildup of fatty tissue—called *plaque*—narrows the passageway of an artery, blood has a hard time getting around the body, supplying oxygen to organs and tissues. And if blood can't make it to the heart, a heart attack occurs. Here's how the stent helps prevent cardiac events and improve circulation.

The tiny mesh stent is collapsed around a catheter that has a balloon at its tip. The catheter is then inserted into the artery, stopping at the blockage.

Inflate the balloon and the stent expands, opening the blocked artery.

Doctors then retract the catheter, leaving the stent in place to prop open the artery and allow blood flow to resume.

OFFICIAL PATENT #9,000,001: ROBO BUDS

No treatise on invention would be complete without a trip to the patent office—if you've gone to the trouble of creating and funding this great new thing that the world needs, you may as well make it official and get the law involved. Getting the government to grant a utility patent is complicated, as you'll soon see, but it all really comes down to persuading a patent professional, called an *examiner*, that your invention meets these three basic requirements.

UTILITY Your invention must be useful, though the established definition of "useful" is very broad. Basically, you can't patent impossible things, like free energy and perpetual motion machines (yes, it is truly a bummer). Utility is mostly a formality; presumably, you wouldn't be bothering with a patent in the first place if your invention had no possible use.

NOVELTY You invention can't have been known to the public before you filed for a patent. Inventions that came before your own are referred to as prior art, and if a search of the patent database or other historical records turns up an invention that is effectively the same as yours, this is called *invalidating prior art*. Establishing novelty requires elbow grease: You'll need to do a thorough patent search to show that your product is related to, but meaningfully different from, specific inventions that have come before.

NONOBVIOUS It must not be obvious to a person skilled in the art (i.e., an expert in the area of your invention). This rule prevents trivial improvements on existing inventions from counting as entirely new inventions. This is the trickiest requirement because you must persuade the examiner that an expert would be impressed or surprised by your results.

[1.] [2.] [3.] [4.] [5.]

106 UNDERSTAND WHAT'S PATENTABLE

No treatise on invention would be complete without a trip to the patent office—if you've gone to the trouble of creating and funding this great new thing that the world needs, you may as well make it official and get the law involved. Getting the government to grant a utility patent is complicated, as you'll soon see, but it all really comes down to persuading a patent professional, called an *examiner*, that your invention meets these three basic requirements.

UTILITY Your invention must be useful, though the established definition of *useful* is very broad. Basically, you can't patent impossible things, like free energy and perpetual motion machines. (Sorry to bring you bad news. You can take off your tinfoil hat now.) Utility is mostly a formality; presumably, you wouldn't be bothering with a patent in the first place if your invention had no physically possible use.

NOVELTY Your invention can't have been publicly known before you filed for a patent. Inventions that came before your own are referred to as *prior art*, and if a search of the patent database or other historical records turns up an invention that is effectively the same as yours, this is called *invalidating prior art*. Establishing novelty requires elbow grease: You'll need to do a thorough patent search to show that your product is related to but meaningfully different from specific inventions that have come before.

NONOBVIOUS It must not be obvious to a person skilled in the art (i.e., an expert in the area of your invention). This rule prevents trivial improvements on existing inventions from counting as entirely new inventions. This is the trickiest requirement because you must persuade the examiner that an expert would be impressed or surprised by your results.

107 KNOW YOUR PATENT TYPES

Every issued patent is given a unique number. There are several subtleties about the numbering system, but generally there are three kinds, which can be identified just by looking at the number.

UTILITY This category is what most people mean when they say "patent." A utility patent covers an invention that does something useful in a unique way. As of late 2016, the U.S. Patent Office had issued about 10 million of them. Just a few of the many famous U.S. utility patents include the elevator (#31,128, filed by Elisha Otis in 1861), the machine gun (#36,836, Richard Gatling in 1862), the light bulb (#223,898, Thomas Edison in 1880), the transistor (#2,569,347, William Shockley in 1951), and the "one-click" online payment system (#5,960,411, Amazon in 1999). These patent numbers use only numerals with no letters.

DESIGN These patents cover only an invention's unique aesthetic aspects, such as shape, color, or detail. As of this writing, there are about 750,000 U.S. design patents, and their numbers are preceded with the letter D. Famous examples include the Statue of Liberty (#D11,023, filed by Frédéric Auguste Bartholdi in 1879), the Coca-Cola bottle (#D105,529, Eugene Kelly in 1937), and the iPad (D504,889, Apple Computer in 2005). Design patents are usually shorter, faster to issue, and less costly than utility patents. Beware of hucksters offering to quickly "patent" your invention for a few hundred dollars—they prey on those who don't know the difference between utility and design patents, leaving them with a patent that protects only how their invention looks.

PLANT Yes, even the patent process allows for a wild card, and the plant patent is it. These patents cover unique varieties of living plants produced by breeding techniques like selection and crossbreeding, and they only apply to plants that can be reproduced asexually (i.e., by cutting or grafting). In the United States, there are currently about 25,000 of them, all with numbers prefixed with "PP."

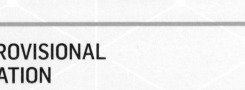

108 STALL WITH A PROVISIONAL PATENT APPLICATION

In the case of competing claims to the same invention, most patent offices award *priority*—that is, award the patent rights—to the inventor who was first to file her application. Because preparing a utility patent application is a time-consuming process, you're allowed to file an abbreviated version, called a *provisional patent application*, which requires only a detailed written description, drawings, and a small fee. From the time the provisional application is received, you have one year to file a regular patent application; if you do so, you may claim the filing date for the provisional application in any dispute that may arise about who was first to file.

109 ENFORCE YOUR PATENT

There are no patent police who go out looking for violations of patent rights. If you get a patent, it'll be up to you to be on the lookout for products and companies who may be infringing. If you find one, it's again up to you to notify them and ask them to stop or, as is more common, to initiate a negotiation about licensing your patent so they can continue their business. You may file a lawsuit in federal court against them at any time, though it's common to try to negotiate an out-of-court settlement first.

110 APPLY FOR A UTILITY PATENT

In general, there are two professionals who can help prepare a patent application: *patent attorneys* and *patent agents*. A patent attorney is licensed to practice law in at least one state and licensed to call himself a patent attorney. A patent agent isn't an attorney but is licensed to assist in the preparation of patent applications. A patent agent can't represent you in court but can help make and submit the application, as well as communicate with the examiner. An attorney, naturally, will cost more than an agent. But with careful study, hard work, patience, and meticulous attention to detail, you can succeed with an application you prepare yourself.

STEP 1 Search for prior art (see #106). It's absolutely in your best interest to discover, as early as possible, if anyone—anywhere, at any time—has patented or published your invention or one close enough to it that yours is no longer patentable. Of all the many professionals you can hire, a search firm is likely the wisest use of your money. If you do the prior art search yourself, put at least 80 hours of work into it.

STEP 2 Construct the claims. These will be used by the examiner (and in any court proceedings that follow) to define, in legal terms, what technology is covered by your patent. You should make the claims as broad as the patent office allows. The catch-22 is that you can't edit your claims after filing, which means you can't narrow them, if they're rejected, without filing a new application and perhaps no longer being first to file. Get an expert to review the claims before submitting.

STEP 3 Make your drawings with black ink on white letter-size paper. (Or build a 3D model in a CAD program and export the line art; see #021–024.) If your invention is abstract—like a process or molecule—you might not need any art apart from 2D flowcharts or diagrams. If it's a manufactured object, you must provide full 3D technical drawings. You'll need at least one general-purpose view that shows the whole invention, plus close-ups of all the features mentioned in your claims. Indicate each feature with a reference number and a leader line that shows its position in the drawing. Label each view with "Fig." Follow it with a number, and include no other text. On the back of each page, put the invention name, your name, and your telephone number.

STEP 4 Write the *specification*—aka, the dictionary for the claims (see #111 at right). Its purpose is to relate the prior art to your invention as described in the claims. Look up some recently published patents to get a sense of the proper style for each section; when in doubt, ask a professional for advice.

STEP 5 Read your application carefully many times, then ask someone you trust to do the same. Confirm the fee schedule and rules for filing immediately before sending, then submit and pay. If you haven't filed many applications and are filing as an individual, you can probably file as a small entity, which reduces the cost. If you opt to file by mail,

you'll have to pay an additional filing fee. Make sure to use express mail and get a receipt from the post office.

STEP 6 The waiting game begins. Within three months, you'll receive an official filing receipt by mail—you can now start using the phrase *Patent Pending*. If you've already gone to market, this notice will deter others who might try to copy and sell your product. (Beware: Marking products with this phrase when you haven't applied for a patent is a crime.) Then settle in for a long slog: The first official response to your application, aka *office action*, can take anywhere from six months to three years. Expect it to include challenges to your claims as well as references to prior art the examiner finds to be invalidating. You must respond by the indicated deadline with a legal argument that addresses the examiner's objections and hopefully persuades her.

STEP 7 Prosecute! The process of pushing an application through the patent office is called *prosecuting*, so don't let that word scare you. A patent agent can do this, or you can do it yourself if you can stay on top of deadlines and seek out qualified advice when you need it.

111 NAIL YOUR PATENT SPECIFICATION

Here's what goes in your patent specification. It helps to look up recently published patents to get a sense of the proper style for each section.

TITLE A simple descriptive phrase that mentions at least one new feature of your invention.

BACKGROUND A one-sentence explanation of the general field or area that your invention relates to.

PRIOR ART A description of the technology that came before your invention, including references to earlier patents and an explanation of their limitations that shows how yours is different.

SUMMARY A brief summary of your invention that paraphrases the claims, using no more than one paragraph for each independent claim.

DRAWING FIGURES A numbered list that identifies, in one sentence each, the views in your drawings.

DETAILED DESCRIPTION A longer section that covers the invention's general structure, then discusses specific parts. Be sure to reference by number and fully explain every numbered part.

OPERATION A longer section that explains the operation of the invention, as opposed to its structure, and describes how each part works during use and what purpose(s) it serves.

CONCLUSION A few paragraphs that summarize new features, talk about other applications, and reframe the invention among the background field.

ABSTRACT If your patent is published, the abstract goes at the start, but you should write it last. In a single paragraph, it should summarize the structure and operation of your invention.

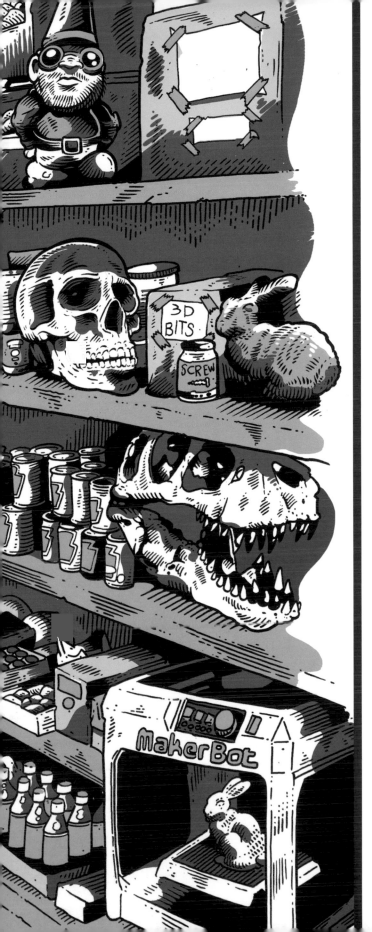

MEET BRE PETTIS

COFOUNDER OF MAKERBOT AND MAKER EXTRAORDINAIRE

When he was just seven years old, Bre Pettis began accompanying his uncle Joe on crack-of-dawn drives around Boston, scouring the streets for junk. The two snatched up any promising finds they came across—furniture, gadgets, bikes, and more—and then sold those reclaimed treasures at weekend flea markets, usually after some requisite fixing up. The tinkering sessions with his uncle—creating a working bike out of the parts of two broken ones, for example—were always Pettis's favorite part. "I loved the thrill of fixing and creating things," he says. "With uncle Joe, I inherited this confidence in taking things apart."

Indeed, Pettis is now one of the most well-known and prolific makers in the United States. His résumé includes stints at Jim Henson's Creature Shop; video blogging prior to the days of YouTube; teaching art; publishing some 200 how-to videos for *Make:* magazine, ranging from a portrait-sketching machine to a T-shirt cannon; creating the popular New York City hackerspace NYC Resistor; and, most recent, becoming a professional entrepreneur and company founder. "For me, the excitement has always been in coming up with something fresh and new that hasn't existed before," he says. "I get a creative rush out of that."

All of his ventures also entail some degree of outreach. Just as his uncle did for him, Pettis is committed to inspiring everyone he encounters to unleash their inner maker. As he says, "A big part of my life is about empowering people to be creative and feel that power for themselves."

Pettis moved from Seattle to New York City in 2007, wishing to be closer to what he perceived as a hub of creativity and innovation. By 2009, he and two partners had founded MakerBot Industries, a Brooklyn-based company that creates affordable

continued on next page

3D-printing tools and now employs some 400 people. "One of the best ways to innovate is to enter a mature market and find different ways of doing it," Pettis says. "We didn't know if anyone would want our 3D printers, but when we launched we sold out instantly."

Pettis's corporate move caused him to catch some heat from the open-source maker community. But Pettis points out that patents are a necessary component of invention should creators choose to go pro. Before filing his own patents, Pettis spent three months reading every Apple patent available and then applied the same principles he gleaned from the computing documents to 3D printing. In the end, he filed around 30 patents for everything from 3D printers for creating infinite-length objects to ones designed with massive, industrial-scale creations in mind. "With regard to patents—especially for technology companies—it's best to enter into a Cold War–type scenario so that you have bargaining chips if things go legal," he says. "Patents give you the freedom to operate, so I came up with as many variations as possible."

In 2013, Stratasys acquired MakerBot for a grand sum of US$403 million—one of the largest tech company exits New York City has ever seen. But last year, Pettis decided it was time to move on. "When I stepped down, I felt like I'd achieved everything I wanted to achieve and that MakerBot was in good hands," he says.

Now Pettis is launching an independent product development workshop called Bre & Co. to create special, handmade-seeming gifts with advanced manufacturing and craftsmanship. "I like to make things, and I want more friendship in my life and the world," he says. "Those are the parameters for this new, sustainable business, where I hope to work with ultratalented people to make beautiful things in service to friendship."

Q: What's your favorite tool?

A: In my new venture, Bre & Co., we use advanced manufacturing to craft heirloom-quality things. Each one gets engraved with a powerful fiber laser that is used in the jewelry industry to engrave metals. The sound of an operator yelling "Fire the laser!" does not get old. This is my favorite tool because it personalizes each thing we make so it's unique.

Q: What's your earliest memory of tinkering?

A: Early on, my parents took me to thrift stores to buy equipment solely for the purpose of taking it apart. Record players and other contraptions were satisfyingly reduced to their component parts. There is a very powerful creative power in deconstruction.

Q: If you could go back through your invention process, what would you do differently?

A: Something I do now was inspired by my grandpa, who worked on the Manhattan Project and later manufactured automated check weighers for assembly lines. He would do what I call "walk-around management," in which he would regularly go around and talk to everyone about what they were working on and their challenges. I wish I could go back in time and do this more in my previous businesses. These days, I've baked it into my schedule.

Q: What happens when you get stuck working on an invention? How do you hit refresh?

A: I helped bring a lot of 3D printers into the world, and so my team is obsessed with experimental, exploratory iterative design. We don't often get stuck; instead, we have the problem of having too many ideas to prioritize and too many opportunities for creativity. Good problem to have!

3D printing is hardly a new kid on the block—the technology has been around since the 1980s, long before the replicator craze hit in the late 2000s. So what's so special about MakerBot? The first kit versions were about as easy to put together as a piece of IKEA furniture, lowering the bar for entry so that the uninitiated could get their hands dirty.

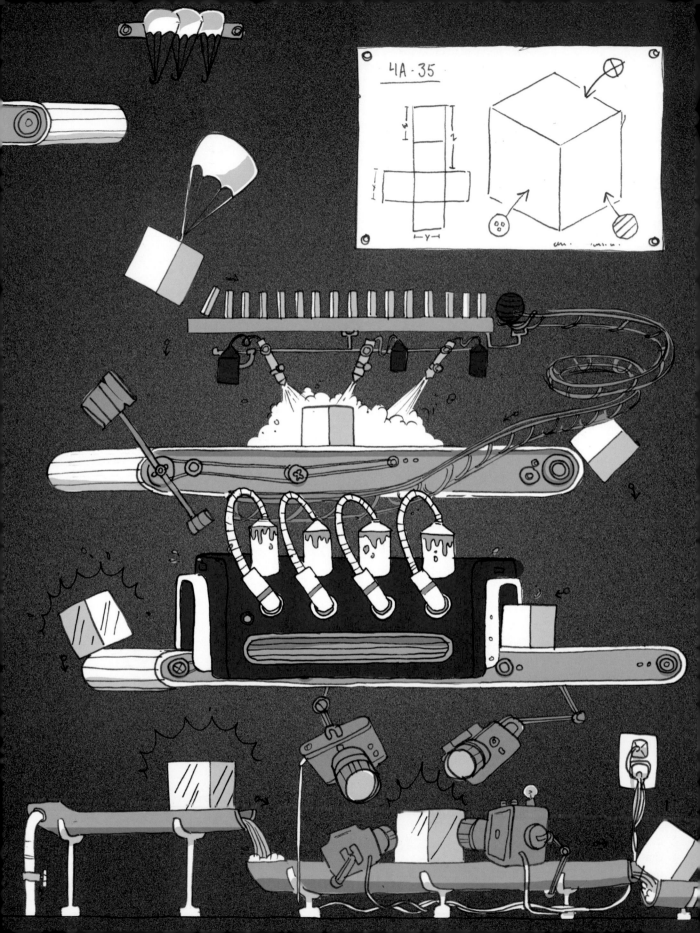

MAKING IT PRETTY

So you've got a working prototype, money in your pocket, and a business well under way. Now it's time to start thinking about what your final product—the one customers actually take out of the box—will look and feel like. In today's marketplace, with its seemingly endless array of options, your product's design is likely to be the first and most crucial element that makes it rise above the competition.

:::

113 SHOW SOME ID

Industrial design (ID) is the art, science, and profession of designing products for mass manufacture. A good industrial designer knows how to take a prototype and optimize its physical form—its shape, assembly, materials, unit cost, and user experience—to the mutual benefit of the people who will buy it and those who will manufacture and sell it.

In many ways, an industrial designer is like a movie director—she doesn't write the story, run the camera, read the lines, or edit the film, but she may well be the single most important person involved in creating the final product. She has a vision for what the product will be like in stores and in the hands of users, and hopefully she has the practical knowledge needed to work with all the players—engineers, manufacturing experts, accountants, marketing types—to deliver that vision on time and at a price point that makes everybody happy.

114 UNDERSTAND THE IMPORTANCE OF STYLING

In perhaps no other field does the fuzziness of the line between engineering and art cause as much confusion as ID. Today, many industrial designers are trained to think of themselves as visual artists first, and many of the world's top ID programs are run by prestigious art schools. There's even been a movement to integrate more technical training, with some institutions offering mixed design and engineering degrees.

But there are various schools of thought, and the words *industrial designer* don't always mean the same thing to everyone. Depending on his education, experience, and personal philosophy, an industrial designer may have lots of, some, or very little interest in the practical side of manufacturing. He may believe that his only job should be to design the look of a product, and that the details of how it's going to work

and get made are not his concern. And he may be right.

In competitive markets, where the consumer has many models to choose from, styling often separates winners from losers. If you're shopping for, say, toasters, you have dozens of comparably priced models to consider, and they all do the same thing. So unless you're one of those rare *Consumer Reports* types who compares features and prices, and crunches numbers to come up with the best value, you're going to make your decision intuitively based on what the different models look like (i.e., whichever toaster looks faster). Having an edge in styling, in being the first of many options to attract the eye, becomes hugely important.

115 GET INSPIRED BY CLASSIC DESIGNS

It's no easy feat to create products that are as appealing to the eye as they are to the hand—that users can pick up and make work without needing instructions. Here are a few outstanding examples to study and learn from.

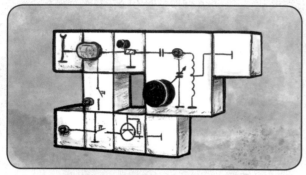

BRAUN LECTRON SYSTEM Developed by Dieter Rams and Jurgen Greubel in 1967, this system of magnetic circuit parts served as an elegant teaching tool.

TRUB PHONE Made of solid rosewood with gold metal buttons, this 1970s beaut by Swiss design firm Gfeller is cut so that the grain of receiver and body align.

IMAC G3 Introduced in 1992, this computer's bubbly shape, bright colors, and clear case made electronics more personable than the usual beige or black boxes.

EAMES CHAIR Ray and Charles Eames broke the mold in the '40s with their iconic, lightweight, and affordable bent plywood chairs for Herman Miller.

CHEMEX COFFEEMAKER Created by Peter Schlumbohm in 1941, this glass beaker streamlined the act of brewing coffee with a wooden collar to shield hands from heat.

ANGLEPOISE LAMP George Carwardine designed auto suspensions before creating this balanced-arm task light. Its springs allow for a wide range of positions.

116 REMEMBER USABILITY

Besides the look of your product, you need to be very concerned with how easy it is to use. In the case of physical products, they must be shaped so that they're comfortable to hold and manipulate, possibly for long periods of time. Controls should be easy to find and operate, and their functions should be made as clear as possible by their shape, placement, and orientation; ideally you shouldn't have to resort to identifying controls with words.

If your product has a software user interface (see #072), it should be easy to learn, easy to remember, and quick to get things done. It should also be fun, or at least pleasant, to use. And it should do all of that without limiting the customer's ability to make the product do exactly what she wants. If your product is software only, you need to achieve all of these objectives not just on one platform but on all types of devices—phones, tablets, desktop computers, video game consoles—that consumers might use to interact with it.

Traditionally, the work of optimizing physical hardware for use by human beings is known as *ergonomics* or *human factors design*, while the analogous process for software systems is called *usability engineering.* If you're venturing into the fast-paced world of consumer electronics, you're going to need a bit of both.

117 OPT FOR FUNCTIONALISM

As an inventor, you'll hopefully be creating a new market, rather than trying to break into an established one, when you introduce your product. If yours is the only game in town, you don't have to worry so much—especially in the beginning—about being the first among many similar products to attract a customer's eye. You are thus free in your ID approach, to skip the flashy styling and instead focus on usability and ergonomics.

Since 1955, world-renowned German industrial designer Dieter Rams has championed a philosophy known as *functionalism*, which advocates exactly this plan of attack. Rather than emphasizing superficial qualities, functionalist design strives to be more substantive and less subject to fashions and trends. In the 1970s, Rams spelled out the principles of functionalism in this set of ten commandments.

- Good design is innovative.
- Good design makes a product useful.
- Good design is aesthetic.
- Good design makes a product understandable.
- Good design is unobtrusive.
- Good design is honest.
- Good design is long-lasting.
- Good design is thorough, down to the last detail.
- Good design is environmentally friendly.
- Good design is as little design as possible.

Functionalism may serve some products better than others, and it'll never be possible to create good design just by following its rules. Nonetheless, Rams's principles stand as handy guideposts for newcomers to the field of ID who may find themselves faced with important decisions about how a product is going to look and feel. They're unlikely to steer you wrong.

118 GET TO THE HOLY GRAIL: THE WORKS-LIKE-LOOKS-LIKE PROTOTYPE

When you begin the process of industrial design, you should already have a working prototype. It may look like something MacGyver built and may not necessarily be convenient or enjoyable to use, but it should have all the functional properties of the finished product. This is called a *works-like prototype*, and it'll be the thing that you hand off to your industrial designer in your first meeting. The process of working with an industrial designer may produce one or more *looks-like prototypes*, which can have all, some, or none of the functionality of the works-like version. The ultimate goal of the ID process is to arrive at a finished *works-like-looks-like prototype*, which has the exact same appearance and functionality that the mass-produced version will have and is put together in exactly the same way with all the same parts in all the same shapes.

To produce this prototype, your ID team will have to use expensive manual craftsmanship or rapid prototyping techniques to create custom parts that, in the manufactured version, will be made cheaply using industrial processes like injection molding (see #164–167). As a result, your prototypes may be very pricey; unit costs in the thousands of dollars are not uncommon. Nonetheless, it's best to insist on at least one complete works-like-looks-like prototype and to spend as much time as possible testing it before you start manufacturing. Expensive though it may be, it'll always be cheaper to fix problems in a prototype at this stage than to fix them in a running factory.

119 PRACTICE HUMAN-CENTERED DESIGN

In 1988, Donald A. Norman published *The Design of Everyday Things*, calling for an integrated approach to product design that puts user experience ahead of everything else. In its pages, he identifies five fundamental principles that will both simplify and revolutionize the way you think about interacting with products—including your own.

AFFORDANCES The relationship between an object's features and a user's abilities acts as a visual shorthand for implied use, helping users intuit what the object is for and how it works. A handle, like on a suitcase, affords grasping and pulling. Buttons afford pushing, knobs afford turning, a touchscreen affords touching.

SIGNIFIERS Designers build these features into an object to explicitly communicate its purpose and operation. Consider the good old *Far Side* comic in which a kid attempts to push open a door clearly labeled "Pull." In this situation, the handle affords grasping and pulling, and the sign (or the signifier) backs it up for those who, well, may not be so swift at picking up on the affordance suggested by the handle.

MAPPING This term refers to the way the layout of a display or a control panel reflects the operation it controls. The classic bad example here is the kitchen stove: The dials are usually laid out in a line, while the burners are arranged in a square. Without labels, how do you know which dial controls which burner? A better mapping would position the dials in a square too—this way, the control panel itself becomes a map of the stovetop.

FEEDBACK If you're going to the trouble to try out a device, shouldn't it at least let you know if it's working? This is called *feedback*, and it needs to be immediate and unambiguous. How many times have you clicked an icon only to have the device do nothing? Too much feedback can also be annoying, so users should be able to control how much they get. (A teachable moment: Clippy, an anthropomorphized paper clip "helper" that constantly popped up to assist with the most basic of tasks in Microsoft Office. Released in 1998 and discontinued in 2003, Clippy is still mocked today.)

CONCEPTUAL MODELS These are simple metaphors that aid users in understanding how products work. Think about the folder icons on a computer desktop—there are no real folders or furniture involved, but the analogy helps people use the computer without getting distracted by technological details.

120 ENHANCE YOUR ERGONOMICS

Expect humans to use your invention? Then you'd better design it for their bodies. This entails applying *ergonomics*—the science of how we interact with products or systems—to minimize discomfort, strain, and ultimately injury . . . whether your creation is a bread maker or a dog-ball thrower.

STEP 1 Collect measurements for your target audience. *Anthropometry* is the science of measuring body sizes (like standing overhead reach, standing eye height, and knee height) in a given population. To ensure fit for all, try designing for the 5th percentile of female users through the 95th percentile of male users. It's hard to measure this many people; a database like ANSUR or CAESAR can help.

STEP 2 Think about how the end user will physically interact with your product and how your design might facilitate reach, sight, and body position. Ask yourself, does the user have to extend her limbs in an uncomfortable manner? If your invention is handheld, how will the user grasp it? Can someone with less-than-perfect vision use it?

STEP 3 Pay special attention to any highly repetitive motions or sustained static postures associated with your product, as well as to the need to hold heavy objects or exert too much pressure during use. Over time, these actions can lead to long-term disability.

STEP 4 Watch your demographic use your and similar products, either in the wild or a focus group (see #060). What environments do they handle the products in? How often? What is their body position during use? The goal here is to understand people's habits, capabilities, limitations, and motivations so you can make the best product possible.

121 CONSIDER CMF

That's short for *color, material, finish*—the process by which industrial designers use market data and manufacturing moxie to select a product's hue, stock, and surface texture. This job is a trend tracker's dream, as it requires steady attention to design currents in fashion, interiors, and consumer goods to advise using, say, a pearlescent instead of a glossy finish. Beyond being on brand and fashionable, there are practical concerns; for example, you can push a smooth button more quickly than a heavily textured one.

Samples make the world of CMF go 'round. When working up a looks-like prototype, ask potential manufacturers for specimens of any metal alloys, plastic resins, molded textures, and painted finishes that you'd like to explore. There are also consultants who forecast CMF trends for consumer products.

122 DISCOVER THE PANTONE SYSTEM

How do you ensure that a factory halfway across the world nails your desired shade of red? With the Pantone Matching System (PMS), the globally recognized color standard. Pantone got started when the printing company M & J Levine Advertising hired a young chemist, Lawrence Herbert, as a color matcher. He took their stock down from 60 to 12 pigments so they could be mixed to make a full array of colors. Six years later, he bought Pantone for US$50,000 and made a basic ten-color palette to license to ink producers. It took off, and X-Rite, makers of color measurement equipment, bought Pantone for US$180 million in 2007.

Today, Pantone employs a 1,757-color palette. Designers of all stripes use its iconic Pantone Guides—thin books of color swatches that fan out—to find just the right shade.

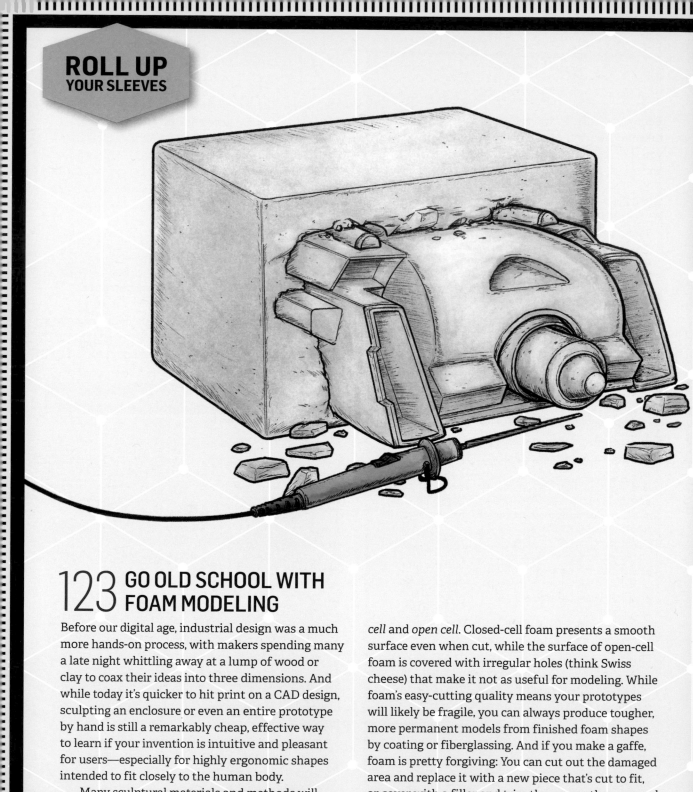

123 GO OLD SCHOOL WITH FOAM MODELING

Before our digital age, industrial design was a much more hands-on process, with makers spending many a late night whittling away at a lump of wood or clay to coax their ideas into three dimensions. And while today it's quicker to hit print on a CAD design, sculpting an enclosure or even an entire prototype by hand is still a remarkably cheap, effective way to learn if your invention is intuitive and pleasant for users—especially for highly ergonomic shapes intended to fit closely to the human body.

Many sculptural materials and methods will work for manual prototyping, but foam is a popular choice because it's cheap and easy to find, glue, and manipulate. Regardless of chemical composition, there are basically two different types of foam: *closed cell* and *open cell*. Closed-cell foam presents a smooth surface even when cut, while the surface of open-cell foam is covered with irregular holes (think Swiss cheese) that make it not as useful for modeling. While foam's easy-cutting quality means your prototypes will likely be fragile, you can always produce tougher, more permanent models from finished foam shapes by coating or fiberglassing. And if you make a gaffe, foam is pretty forgiving: You can cut out the damaged area and replace it with a new piece that's cut to fit, or cover with a filler and trim the excess, then conceal the evidence with paint or another coating.

When you're cutting up foam, be mindful of the dust; those tiny particles can be hard on the lungs. Work in a ventilated space or wear a dust mask.

EXPLORE DIFFERENT MODELING MATERIALS

Before you start trying to repurpose the nearest takeout container in your foam looks-like prototype, you should know that not all foam is created equal. You need a material that will cut smoothly, hold fine details, and take glue and paint without being damaged by them. Here are the two most popular modeling foams, and some chemicals that work well with them.

POLYSTYRENE FOAM (STYROFOAM)

The familiar white beady stuff that makes up coffee cups, cheap coolers, and packaging inserts is called *expanded polystyrene (EPS) foam*. It cuts well with a hot wire, but its large-grained structure makes it hard to carve smoothly with cold tools. Instead, look for the blue stuff sold in sheets at the hardware store as building insulation. Called *extruded polystyrene (XPS) foam*, it is heavier, has a finer structure, and takes smoother cuts from hot and cold edges.

PHENOLIC FOAM (BALSA-FOAM)

In the same chemical family as florist's foam, this stuff is usually orange or yellow. Specifically designed for carving, Balsa-Foam is easy to find in thick blocks—unlike XPS, which is only available in thin sheets that you have to layer up. It also takes detail better and is more chemically resistant, so you can coat or paint it without worrying about the solvent eating the foam. Still, I recommend testing a new coating on scrap before applying it to your hand-sculpted masterpiece.

AUTO-BODY FILLER (BONDO)

A two-part mixture with a pink or gray appearance, Bondo hit the auto-body scene as a pre-paint-job filler for small dents and scratches. It's tough, easy to sculpt, and widely available, and everyone from modelers to special-effects artists to—yes—industrial designers has appropriated it for a wide variety of off-label uses. Warning: Body filler is oil based and cannot be directly applied to polystyrene—it'll dissolve the plastic. So cover your polystyrene with spackle or another water-based filler before applying body filler.

FORMEROL (SUGRU)

Sold in small, airtight, single-serving packets (kind of like fast-food ketchup), Sugru is a no-mix, air-curing silicone rubber that you can freely sculpt for about 30 minutes after opening. When set, it has a bouncy, elastic texture that's perfect for modeling ergonomic elements like handgrips and buttons, as well as protective pieces like bumpers and feet. It comes in a number of colors and has a shelf life of one year, so it's best to buy in small quantities.

STYRENE (ABS)

An acronym for *acrylonitrile-butadiene styrene*, ABS is the plastic you find in LEGO bricks and model kits. Most hobby shops sell it as a solid white material packaged in sheets, strips, tubes, and more. You can bond this plastic easily to itself with plastic model cement or other common solvents. Tip: Superglue works well for attaching ABS parts to other materials, especially if the bonding surface is roughed up a bit with sandpaper first.

CYANOACRYLATE (SUPERGLUE)

This tenacious adhesive is colorless, transparent, and available in thicknesses ranging from watery to that of Jell-O. It sets within seconds, creating strong bonds between dissimilar materials and textures. Not all formulations are okay to use on foam; when in doubt, opt for a formulation marked "foam-safe." When working with superglue, keep a tube of debonder handy to free trapped fingers and other accidental adhesions.

125 SCULPT A MOCKUP IN FOAM

Now let's model a simple shape in closed-cell foam—you'll end up with a durable looks-like prototype coated in tough auto-body filler. To get started, you'll need a few tools: a fine-point marker, scissors, a Zona saw or fine-tooth hacksaw blade, a hot-wire cutter, fine and coarse files, a hobby knife, an old toothbrush, a popsicle stick, and a small paintbrush. You'll also need the following materials: drawing paper, blue XPS Styrofoam or Balsa-Foam, spackle (or other water-based filler), PVA glue, Bondo (or other oil-based auto-body filler), wet-or-dry sandpaper (in 200, 400, 600, and 1,000 grit), a sheet of styrene plastic, superglue, commercial spray paints, and Sugru (optional).

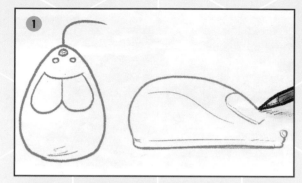

STEP 1 Draw what you're going to sculpt, refining a series of thumbnail sketches (see #017). Once you have the shape in your mind's eye, make a series of plan views from the front, top, and side.

STEP 2 Prepare a slab of foam in your required size either by cutting down a larger piece or by stacking and gluing smaller pieces. Transfer your plan views to the front, top, side, and possibly other surfaces of the "blank"—you can cut out paper templates and paste them on, or just draw the profiles right on the foam.

STEP 3 Time to sculpt. First, use a saw to rough out your shape, cutting the profiles that you transferred to the blank. (If you're working with Styrofoam, use a hot-wire cutter instead.) Next, round over the edges and corners with a coarse file.

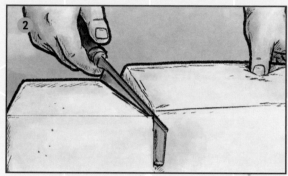

STEP 4 Cut out details using your hobby knife and fine abrasive tools, such as needle files. When cutting foam with a knife, try pushing and sliding as if it were a loaf of bread. Otherwise, the foam may tear off in chunks, leaving a pockmarked surface.

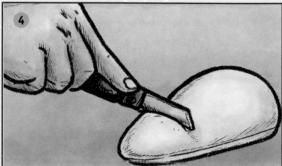

STEP 5 File so the foam is smooth and free of large imperfections. Clean dust out of the pores with an old toothbrush, then apply an even layer of water-based interior filler (like spackle) using a popsicle stick. This application will protect your foam from the auto-body filler that we'll layer on in the next step. (If you're using Balsa-Foam, you may be able to skip this water-based filler, but test on a piece of scrap to be sure.)

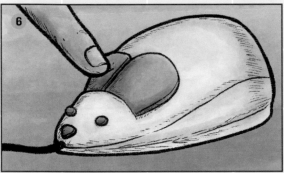

Let the spackle dry overnight, then smooth out any bumps with fine sandpaper. Brush on a coat of PVA glue; let that dry overnight too.

STEP 6 Mix up some auto-body filler and apply a layer to your creation. Once it's dry, rub out ridges or bumps with a fine-tooth file, then sand the whole thing with four different sandpapers in ascending grit (200, 400, 600, and 1,000). You're finally ready for the fun stuff: Cut any surface details (charmingly called *greeblies* by those in the biz) from a styrene sheet, then apply them with superglue before priming, painting, and sealing using commercial spray paints. You can also add rubbery (aka *elastomeric*) texture or details—for instance handgrips, bumpers, or feet—by applying and sculpting Sugru, an air-curing, silicone-rubber modeling compound.

126 CRAFT DETAILS LIKE A SCALE MODELER

When it comes to sculpting small parts, nobody knows more than the folks who build miniature models, whether of cars, planes, militaria, or sci-fi stuff, especially those who are into *scratch-building*, which means starting from stock materials rather than preformed kits.

For working in plastic, styrene is the material of choice. The best way to cut sheet styrene is to score a groove along a straightedge with the back side of a hobby knife, then bend along the score line until it snaps. You can cut interior openings by first piercing the area to be removed with a drill, then enlarging the hole

with a file, saw, or *nibbler*, a tool that bites out a little chunk every time you squeeze the handles.

You can also shape styrene with heat, such as a candle flame. With practice, you'll be able to soften a piece of rod and stretch it out to your desired diameter. A step up is the *strip heater*, a benchtop tool that applies heat along a straight line and makes it easy to bend sheet plastic. (The bend sets hard when it cools.) Finally, on the pricey side, there's vacuum-forming, in which an entire sheet of plastic is heat-softened, then pulled into shape against a master or "buck" form by drawing the air out from between them.

127
MEET AYAH BDEIR
FOUNDER AND CEO OF LITTLEBITS

Ayah Bdeir has always had a knack for creativity. Born in Montreal and raised in Beirut, she dreamed of becoming an architect. She also excelled at math and science, and had a love for taking things apart. Noting this with interest, Bdeir's parents and teachers encouraged her to pursue engineering instead of architecture, and when it came time for college she dutifully complied. But she found the coursework to be dry and boring. It wasn't until she enrolled in MIT's Media Lab for graduate school that she realized engineering might be the right fit for her after all. At MIT's program, art, technology, design, multimedia, and the humanities converged, providing Bdeir with her first glimpse into "how you can combine amazing advances in engineering with great ideas of design and social change."

In 2006, the year she graduated, Bdeir made her first invention. Called *Random Search*, it was an electronic art piece that consisted of a wearable, reactive undergarment that recorded, shared, and analyzed the experience of invasive airport searches.

After graduating, though, Bdeir had no plans for starting a company. She focused on creating artwork that used technology to express ideas around identity, culture, and tradition. She was also interested in outreach and developed an early version of littleBits, an open-source platform of easy-to-use electronic building blocks, as part of a project to help industrial designers improve their prototyping process. LittleBits was also meant to make it easier for non-engineers to incorporate electronics into their work. Bdeir was shocked at the reception, however. "As I started demoing littleBits, there was enormous excitement from kids, parents, and teachers who wanted to use littleBits in ways I had never anticipated," she says. "I realized this was a very big opportunity

continued on next page

to change the way kids learn about technology and to change their relationship with technology from passive consumers to creative problem solvers."

Bdeir followed her hunch and founded littleBits Electronics in 2011. Harkening back to her creative roots, design—not just functionality—was very important to the company. Beautiful products, she adamantly believes, are better products. "I made art and design the center of our product and our company," she says. "Our second hire was a designer." As a result, littleBits's electronic building blocks are outfitted in a variety of attractive, bright colors, combining to create multicolored rainbows when they're snapped together. "Adding beautiful design is another way in which we're making our product more accessible, by making them more appealing to everyone," she says. "It helps invite people outside the choir and helps humanize products."

Indeed, reaching as many people as possible is one of Bdeir's goals. "Our mission is to unleash the inventor in everyone," she says. Doing so, she believes, will result in more innovation, more creativity, and better products, rather than the usual mass-market-designed stuff that "we just purchase and throw out." She seems well on the way to fulfilling that ambition. LittleBits is now used in more than 3,000 schools and has shipped to more than 150 countries around the world. The company also employs more than 100 people and is supported by a global community of inventors and makers, including children. For example, an 11-year-old boy named Vedant has created a whole slew of amazing projects, like a penny-operated car and a robot that blows bubbles.

"In order to create solutions for 21st-century challenges—economic, environmental, medical, and more—we need a diverse pool of inventors and designers and problem solvers," Bdeir says. "I go on Instagram and see what people make with littleBits, and I'm so proud."

Q: What's your favorite tool?

A: I love working with cardboard for superquick prototypes, so the tool I usually reach for is a box cutter.

Q: Are there any other inventors or makers who particularly inspire you?

A: I'm inspired by all makers, especially the younger ones. We have an 11-year-old girl in our community named Anahit who is teaching herself to code with Arduino and prototyping amazing inventions with littleBits, like an echolocator device for the vision impaired. It blows my mind.

Q: What do you do when you get stuck working on an invention? How do you hit refresh?

A: I like to bring in a fresh point of view—someone from a completely different team. It helps to surround yourself with problem solvers. I also browse our invention page; there are so many incredible ideas there.

Q: What was your greatest challenge in making your invention a reality?

A: It was very, very hard at the beginning on the manufacturing side, especially before I had funding. Manufacturers would ask me, "How many units do you want to make?" and I would say "100." They would ask, "100,000?" and I'd say, "No, just 100," and that ended the conversation. I had to do a lot of research and courting before I found someone to make a prototype and go into production with us.

Q: If you could go back through your invention process, what would you do differently?

A: I would probably have introduced the product to the general market sooner—I spent several years iterating on the design before I launched the company.

BARGRAPH

BUTTON

DIMMER

POWER

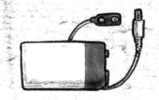

SERVO

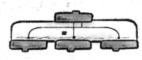

FORK

ARDUINO

BATTERY

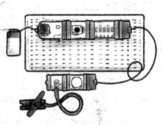

MOUNTING BOARDS

These bright, color-coded electronics promise that your first circuits will be fun—and maybe even easy—to make. Choose from power sources (blue); inputs like motion triggers, light sensors, and keyboards (pink); extensions such as wires and specific boards for Arduino and MIDI (orange); and outputs like motors, speakers, fans, and more (green). Once you've selected your components, you skip the usual soldering and snap together their magnetized ends to see the current flow. The company also provides clever kits so users can swiftly try their hands at making more elaborate objects, like the car you see here.

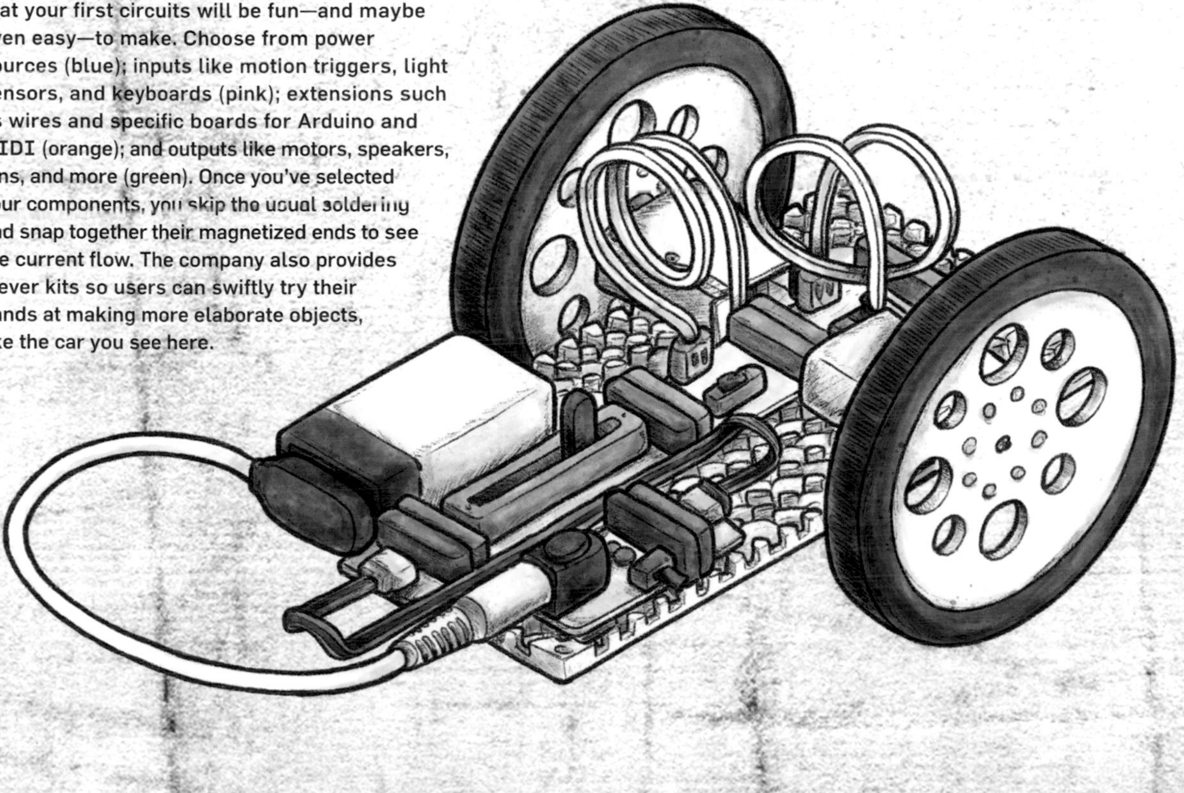

128 GET INSPIRED BY LEADERS IN DESIGN

For some, it's not "form follows function"—it's form *is* function. Here are noted thinkers on the value—and signs—of good design.

"PEOPLE THINK THAT DESIGN IS STYLING. DESIGN IS NOT STYLE. IT'S NOT ABOUT GIVING SHAPE TO THE SHELL AND NOT GIVING A DAMN ABOUT THE GUTS. GOOD DESIGN IS A RENAIS-SANCE ATTITUDE THAT COMBINES TECHNOLOGY, COGNITIVE SCIENCE, HUMAN NEED, AND BEAUTY TO PRODUCE SOMETHING THAT THE WORLD DIDN'T KNOW IT WAS MISSING." *– Paola Antonelli, architect and director of R&D at the Museum of Modern Art*

"When you're doing something that is distinctive in design from the outset, it's very clear who is the original and who is the copycat." *– Gadi Amit, Fitbit designer*

"RECOGNIZING THE NEED IS THE PRIMARY CONDITION FOR DESIGN." *– Charles Eames, renowned industrial and furniture designer*

"An aircraft that looks beautiful will fly the same way." – Kelly Johnson, Lockheed Martin engineer famous for his contributions to the SR-71 "Blackbird"

"We never start with this nice-looking picture of our vision, of how something should look at the end … At every iteration of the prototyping, we try to update 'What would the product look like?' as if this would be the final stage of development. And during this iterative process, the details get carved out." *– Carol Zwick, inventor of the award-winning Mirra chair*

"GOOD DESIGN KEEPS THE USER HAPPY, THE MANUFACTURER IN THE BLACK, AND THE AESTHETE UNOFFENDED."
– Raymond Loewy, designer of NASA's Skylab Space Station, the Coca-Cola vending machine, Studebaker Avanti, Schick electric razor, and more

"DESIGN IS NOT ART. DESIGN IS UTILITARIAN, ART IS NOT."
– Massimo Vignelli, designer of the iconic New York City Subway maps

"To this day, I cannot tell whether my dog is interested in the bone-shaped biscuit because it fools him as such, or whether, after my shaping the biscuit in an effort to cater to his taste, he feels duty bound to fool his master by simulating an interest in it."
– Carleton Ellis, designer of the bone-shaped dog biscuit

"I follow trends religiously. Fashion, architecture, art, design, retail. I try to absorb it all as a way to keep track of what people are interested in, what colors are popular, what people are wearing. But when it comes to my own products, I think about what I would want to own."
– Max Gunawan, inventor of the Lumio light

"INVARIABLY WHEN WE DESIGN SOMETHING THAT CAN BE USED BY THOSE WITH DISABILITIES, WE OFTEN MAKE IT BETTER FOR EVERYONE."
– Donald Norman, author of The Design of Everyday Things

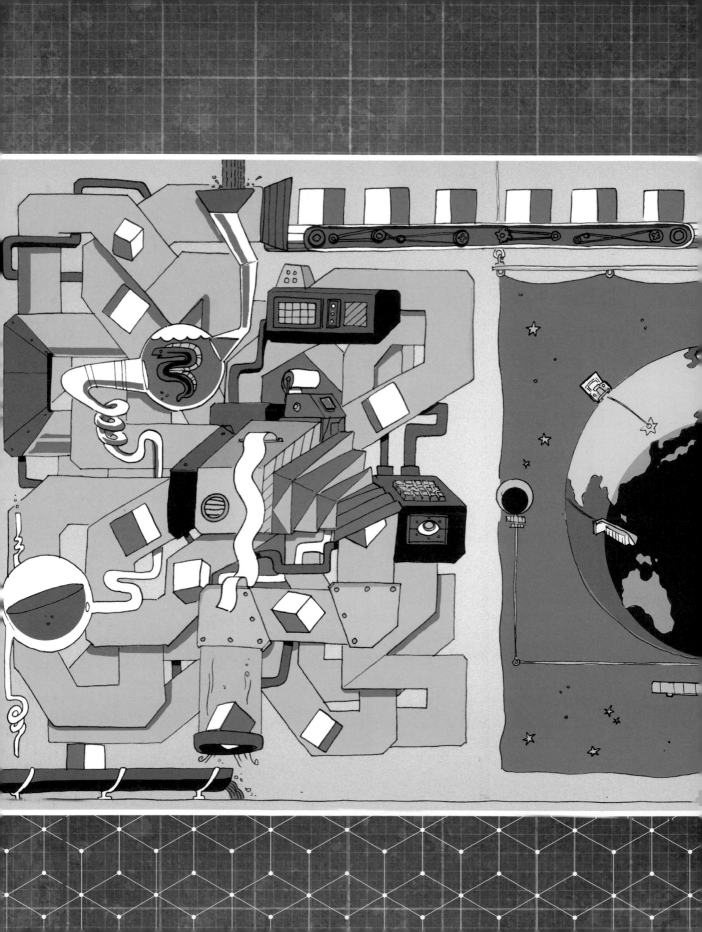

MAKING MANY

DESIGN FOR MANUFACTURE

Most of the magic that makes our modern world happens on a factory floor. The more you think about factories, the more you learn to appreciate the beauty of these building-size machines that we hide on the outskirts of town. Their lesson for inventors? Your product is not just the thing that goes in the box but the process used to put it together.

129 TREAT INDUSTRIAL AND MANUFACTURING DESIGN AS ONE STEP

Prototyping an invention so it works is one thing, and designing it so it's intuitive to use is another. Don't shoot the messenger here, but you're not done yet: You also must finesse the design so it's easy to make, a process known as *design for manufacture* (DFM).

In an ideal world, industrial design and design for manufacture happen simultaneously. Doing rounds of development to arrive at a works-like-looks-like prototype (see #118) will naturally make you consider how to manufacture it—which parts will be injection-molded, which assembly will be glued first, and so on.

In the real world, however, things don't always work out that way. Your manufacturing partners may not be able to execute your product designer's plan as specified (or alternately, they can do it better and more cheaply using another approach). So the earlier you can get the factory involved in your design process, the more time and trouble you'll save. Which means you should be looking for a manufacturing partner at about the same time you're hunting for an industrial designer.

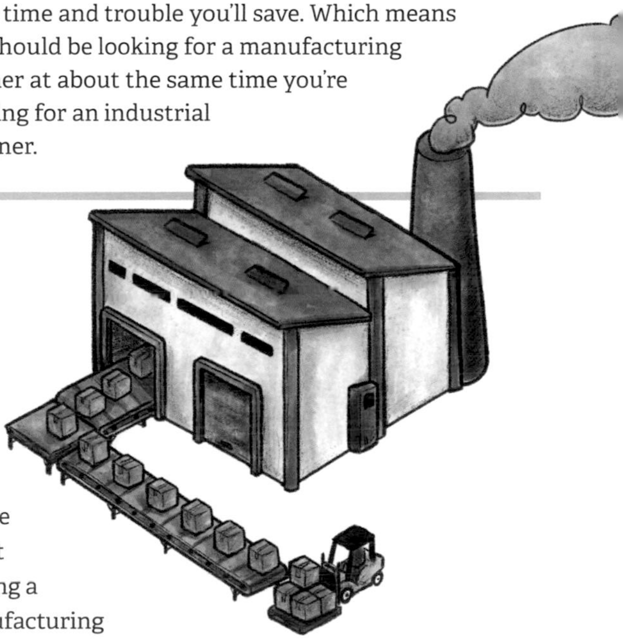

130 LEARN FROM THE LOCALS

Lots of people will tell you that you need to use an overseas factory to keep your labor costs down, but that bit of conventional wisdom overlooks the huge value add of the "just next door" factor, especially during early production runs. If you have to fly literally halfway around the world and deal with people who don't speak the same language every time a problem crops up, the cost savings of foreign labor can evaporate pretty quickly.

The time-honored (if deplorable) approach is to develop your manufacturing process at a local factory, then move production overseas once you've got the bugs ironed out. I would not go so far as to advocate that philosophy, but I will say that the best general advice about finding a manufacturing partner is to start in your own backyard. Who has factories in your hometown? What are they making? Can those people help you or at least teach you best practices? If not, what about the next town over?

131
WATCH *HOW IT'S MADE*

The Discovery Channel TV show *How It's Made* has produced more than 300 fascinating half-hour episodes documenting the inner workings of all kinds of factories, all around the world. Minus commercials, that's more than 115 hours of first-class manufacturing education available at almost no cost. Episodes are regularly aired on major networks and streamed via various online services. In addition, the show's creators make many of its segments free to watch online.

Every show is divided into three seven-minute shorts, each featuring a working production line for a single product ranging from cultured pearls to combine grain harvesters. If you're interested in running an assembly line of your own someday, you should watch this show obsessively, perhaps with your team, and think and talk about what you see. Find the episode featuring the product that's most like the one you want to build, watch how they do things, and ask yourself why? How would you do it differently? Or even better?

132
CONSIDER KITTING

Some products come fully assembled, ready to use right out of the box, while others come in a bundle of pieces, parts, and cheap tools that consumers use to build up the product themselves—aka, a *kit*. From a manufacturer's point of view, the beauty of a kit is that you just source the parts and pack them in a box with good instructions (see #203–207). And from the customers' point of view, building the product on their own gives a sense of pride—plus, they're better prepared to fix it if something goes wrong. The downside, of course, is that "some assembly required" scares off potential buyers who are looking for the instant gratification of unwrapping their purchase, plugging it in, and using it.

The choice to kit or not to kit depends on your market size and demographics. For small markets of people who like to get their hands dirty, kitting is a godsend. For big markets of folks after a quick solution, customers need to see a large savings over preassembled goods before a kit becomes attractive.

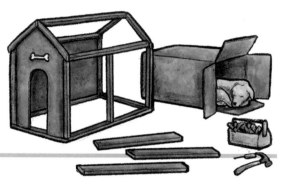

133
FIND A FACTORY

Regardless of whether you stay onshore or hop overseas (see #139), there are many considerations in locating and partnering with a manufacturer.

REACH OUT In our hyperconnected world, it can be surprisingly difficult to find a factory that can do what you're after. If you plan on manufacturing in Asia, pay a visit to Alibaba or IndiaMART, large ecommerce websites where suppliers advertise their services. For stateside manufacturing, try your luck with Maker's Row or ThomasNet. Buyer beware: Scams are common, so always verify the business license, location, ownership status, and certifications of a potential partner, as well as talk on the phone, request photos of the factory, and maybe even hire an agent to pay the company a visit and assess its viability. (Yes, that's a real job.)

ASK THE RIGHT QUESTIONS Once you've found a likely candidate, you'll need to send what's called a *request for quotation* (RFQ). First, ask a few questions to make sure the factory is even in your ballpark. Knowing the *minimum order quantity* (MOQ) upfront will determine if you can afford to do your initial run there, but request pricing at various run sizes so you get a sense of how they discount. Be clear on their payment terms too so you don't get surprised by, say, needing to pay for the entire production run before they hit start, which is indeed how many factories operate.

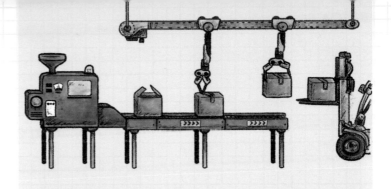

134 MAKE THINGS EASY TO MAKE

Save yourself pain (and cash) by keeping your assembly line in mind from the very beginning of the design process. Here are some basic guidelines.

MAKE FRIENDS Like anything in life, where you have friends you have people who can help you out. Get to know the folks at your factory and establish a positive rapport.

SIMPLIFY EVERYTHING Other factors being equal, always opt for the design that has fewer parts in less complicated shapes.

DESIGN FOR SCALE If you're only building ten units of a product, its design can and should be very different than if you're building 10,000.

USE OFF-THE-SHELF PARTS Every custom part will add tooling costs, including parts and labor. It's usually in your interest to use as few of these as possible.

KEEP TOLERANCES LOOSE Precision is all well and good when it's needed but wasteful when it isn't. Specify tight fit only where clearly justified by functional purposes.

GO WITH FEWER STEPS Design parts so they can be made in a single operation, avoiding follow-on steps (like plating or polishing) if possible. Where multiple steps are required, design the part so it can be moved as little as possible between them.

SPECIFY CHEAPER TOOLS Where you can't eliminate a step, make sure you're doing it with the most economical process. Punching holes is cheaper than drilling them, for example.

MAKE EVERYTHING LIGHTER In general, parts that weigh less use fewer materials and will save you in shipping costs. If you can make something lighter without compromising quality, go for it.

AVOID SHARP CORNERS Whether the part is injection-molded from plastic or stamped from metal, the corners should be as round as possible, both inside and out.

135 TAKE A TRIP DOWN THE ASSEMBLY LINE

In a modern assembly line, the product moves past stationary workers who put it together in steps, with each specialist doing her thing as the line rolls by. Each task is designed to cost a minimal amount of time, energy, and motion. The productivity of a continuous assembly line is limited only by its slowest step—if the longest operation takes four minutes, for instance, the line can produce one complete product every four minutes.

One of the earliest examples of this approach actually dates to Italy at the time of Dante, when Venice was the capital of a wealthy seafaring republic that dominated trade across the Mediterranean. The Venetians produced a vast fleet of warships at the city's famous Arsenal, which, at its peak in the 1600s, could turn out one vessel every day using assembly-line methods, with the ships moving along canals from one workshop to the next. The rest of Europe would have to wait until the Industrial Revolution came up to speed in England to see the factory system emerge. The Portsmouth Block Mills, which produced wooden blocks (i.e., pulleys) for the Royal Navy from 1808 right up until the 1960s, are often cited as the first continuously operating production line.

In the United States, the meatpacking industry in turn-of-the-20th-century Chicago was an early milestone. Around 1910, an engineer at Ford Motor Company visited the city's famous Union Stock Yards, where he saw animal carcasses hanging on an overhead conveyor moving past stationary butchers, each of whom made the same cut over and over again. He took the principle back to Detroit and suggested that the auto tycoons apply it to producing cars. By 1912, that engineer, William "Pa" Klann, was in charge of the Model T assembly line, which by 1914 turned out one new car every 99 minutes. By 1930, every major automobile manufacturer was using an assembly line, and many who had been slow to adapt were out of business.

136 MASTER MANUFACTURING PROCESSES

Most everyday assembly lines use standard tools designed to perform common manufacturing processes, of which there are probably hundreds if not thousands of distinct types. These, however, can mostly be lumped into four big categories.

CUTTING These processes include punching, die-cutting, sawing, torch-cutting, laser-cutting, and all forms of conventional machining, as well as exotic stuff like electrical discharge machining, water-jet cutting, and photoetching. The theme, as in the old joke about carving an elephant, is removing all the stuff that doesn't belong while keeping all the stuff that does. While it may seem natural to cut things right off the bat, in practice the uncut stock often ends up being easier to handle, so cutting operations get pushed back along the line.

FORMING This family of processes includes blow molding, injection molding, vacuum-forming, rotational molding, stamping, bending, swaging, forging, metal casting, glass blowing, steam bending, and composite laminating, just to name a few. Here the idea is to take the material that's already there and mash it around into a new shape, which it then holds. Waste is often recycled—like sprues and runners in injection molding, which get fed straight back into the melt after trimming.

JOINING Such processes include fastening (be it with screws, bolts, rivets, staples, or any other "hold it together" hardware), press-fitting (in which parts are designed to simply interlock and stay that way), gluing, soldering, heat-staking, and welding in all its myriad forms. Here you're putting smaller bits together to make bigger ones. Believe it or not, the assembly line will likely employ as many joining steps as there are parts in the BOM.

FINISHING This is usually the last step in any part's manufacture—you don't want to mess up a fancy finish by cutting, stamping, or welding all over it. Processes include printing, sanding, grinding, polishing, sand- and bead-blasting, engraving, spray painting, electroplating, powder coating, anodizing, galvanizing, and vacuum metalization.

TOOL TIME

137 STAMP PARTS WITH A FOUR-SLIDE MACHINE

A *four-slide machine* (aka a *multislide machine*) pulls metal wire or strip stock off a roll at one end, and spits out complex bent, stamped, or punched parts at the other—think springs, brackets, and hooks. It consists of a table-shape tooling area bounded by rotating shafts on all four sides. The shafts are geared together and driven by a single motor, and they control the movements of mechanisms that work the stock: feed rollers to pull the stock through, a straightener to flatten the curve from the roll, a stamping press to shape the outer edges and interior openings, a cut-off shear to separate single parts, and a set of sliding tools that strike the part against a central forming station from four different directions. A heat treatment is usually needed to set the metal.

A single four-slide can manufacture complex metal parts very quickly—rates of 50,000 per day are common. The slides can operate all together, in sequence, or in combinations, so it's very versatile and easy to reconfigure. It's also not too expensive, as industrial tools go, and makes a cost-effective choice for both low- and high-volume production.

||||||||||||||||||||||||||||||||||||||||||||||||||||||||||||||||||||||||||||||||||||||

138 GEEK OUT ON SPECIALIZED MACHINES

The machines that crank out our world are often standardized so they can handle many different products. Others, however, are truly special snowflakes, designed to make just one thing with utmost efficiency.

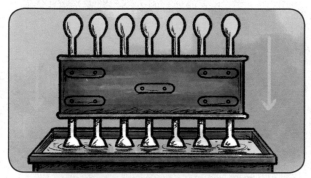

BALLOONS Rotating trays of forms get dipped into vats of dyed latex, then scooted off to a bank of spiraling brushes, which roll up the balloons' bottoms into lips.

BUTTONS A sheet of colored resin runs through a press, which stamps out tiny discs. A variety of cutting heads then carve the discs and punch them with holes.

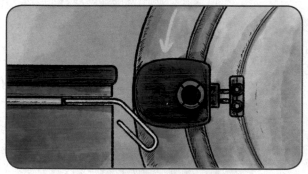

PAPER CLIPS This machine feeds wire through a hole, exposing it to a spinning array of heads that bend the wire in on itself each time one makes contact.

FUSILLI An extruder pushes dough through a die to give it a distinct spiral shape. As the dough comes out, a blade sweeps across and cuts it to the desired length.

ROPE A ring of bobbins revolves around a central spindle. As they go, yarn wraps around the spindle, forming a complex and multicolored braid.

GLOBES Pre-printed, pre-punched globe halves go over molds, which then rise into an embossed press. The resulting domes are fastened together in a sphere.

139 GO LOCAL OR OVERSEAS

Let's just start by saying there's no clear-cut answer for whether to manufacture your product locally or overseas. It all depends on how fast you need it, how much money you have to spend, and how highly specialized the invention is. Technicalities aside, the quality of your relationship with your manufacturer is the most important deciding factor. If you've got a good thing going, better to stick with it if you can afford it.

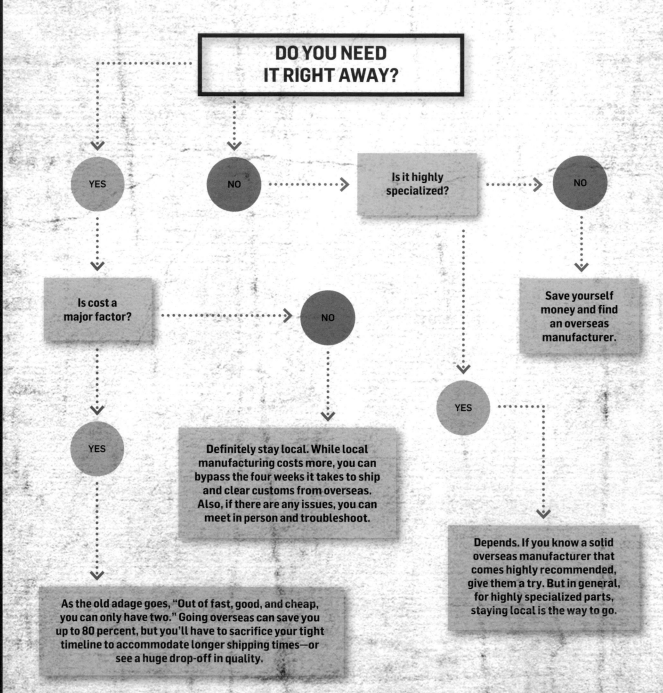

DO YOU NEED IT RIGHT AWAY?

YES

NO

Is it highly specialized?

NO

Is cost a major factor?

NO

Save yourself money and find an overseas manufacturer.

YES

YES

Definitely stay local. While local manufacturing costs more, you can bypass the four weeks it takes to ship and clear customs from overseas. Also, if there are any issues, you can meet in person and troubleshoot.

Depends. If you know a solid overseas manufacturer that comes highly recommended, give them a try. But in general, for highly specialized parts, staying local is the way to go.

As the old adage goes, "Out of fast, good, and cheap, you can only have two." Going overseas can save you up to 80 percent, but you'll have to sacrifice your tight timeline to accommodate longer shipping times—or see a huge drop-off in quality.

140 ASSESS LIFE CYCLE

Going green means taking a look at your invention in what's called a *life-cycle assessment* (LCA). An LCA tries to account for all the energy and materials that go into a product, as well as the emissions that come out of it. This report spans from cradle to grave, covering how your raw materials are extracted, refined, and transported; how your parts are manufactured, assembled, and packaged; and how your finished units are distributed.

Running an LCA starts with a flowchart that documents the inputs and outputs at each production stage. You gather data from your manufacturing partners on every input and output that's associated with your product, including the volume that's created or consumed at each stage. Then each substance receives an impact rating based on how much it contributes to a specific environmental woe. These problems are weighted and ranked according to expert analysis and used to arrive at a profile of your product's environmental effects. From there, it's up to you to make needed adjustments to your product cycle.

141 PRACTICE DESIGN FOR DISASSEMBLY

Design for disassembly (DFD) is exactly what it sounds like: designing products so they're easy to take apart for repair, refurbishment, and recycling. The financial logic of manufacturing naturally leads engineers to focus on rapid, cheap, and durable assembly methods, but many of these techniques also make it hard for the buyer to do anything with the product besides chuck it in a landfill when its useful life is over. For instance, gluing and welding can discourage repair, because parts must be broken apart and are then hard to put back together. And *multishot molding*—in which two or more plastics are permanently molded together—can discourage recycling, because the different plastics may need to go in different streams.

Going to the trouble of making your products easy to take apart may not seem worth it, but consumers are increasingly aware of the environmental impacts of product design, and making irresponsible choices can hurt your brand. Likewise, customers who must replace a broken product just because it's too difficult to repair are going to be twice as upset—and twice as likely to write a scathing review.

Besides providing marketing opportunities, practicing intelligent DFD can reduce costs in many cases. After all, stuff that's easy for unskilled customers to take apart without fancy tools can also usually be put together by cheaper labor in cheaper factories.

142 EMBRACE THE TENETS OF ECODESIGN

To get ahead of potential environmental problems in your product's life cycle, try taking some of these guiding principles to heart during your DFM phase.

KEEP IT CLEAN Work with your manufacturer to source substitutions for any toxic materials or processes required to make your product. Sometimes you just can't get around a real nasty one; in that case, rely on a closed loop system to isolate and recycle the offending ingredient.

INVITE UPGRADES If you're working on an electronic device, prevent obsolescence (read: landfill) by pushing upgrades to your users rather than requiring them to purchase a fresh product.

STAVE OFF WEAR AND TEAR Select surface treatments that will help prevent dirt and corrosion from damaging the product before its time is up.

WATCH YOUR WEIGHT Try to use lightweight materials in order to cut down on fuel needed for transport.

DON'T FRANKENSTEIN PARTS Avoid hard-to-recycle metal alloys and permanently joining materials that require different recycling methods.

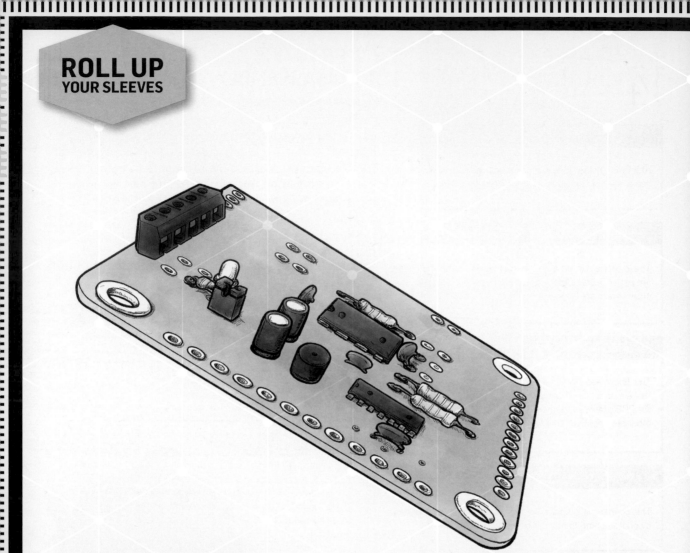

143 GET TO KNOW PCB FABRICATION

Grab a screwdriver and pop open the nearest gadget—you'll likely find one of these things inside. Typically a sheet of green composite covered with a complex pattern of copper lines (called *traces*) and various electronic components, the *printed circuit board* (PCB) is the all-grown-up-and-ready-to-mass-produce version of a circuit that you might first prototype by hand on breadboard (see #049).

The modern multilayer PCB is a miracle of manufacturing technology. Your key to understanding how it works is to think in terms of layers: This thing is basically a sandwich consisting of anywhere from two to several dozen very thin sheets of conductive metal, each with its own pattern of circuit traces separated by layers of insulating fiberglass. To attach

the components to the circuit, you either run the components' wires (called *leads*) through holes in the PCB and solder them on the underside, or you use *surface-mount components* with tiny legs or pads that can be soldered directly to the traces.

While something that looks so complicated may seem intimidating to make yourself, it's actually pretty easy. These days, PCB design software is cheaply or even freely available, and very user friendly. And the physical PCB fabrication process is so streamlined and efficient that it's possible to order custom-manufactured boards, in quantities as small as one, for just a few dollars. Some companies will even solder on the components for you, but that tends to take a bit more coin.

144 LEARN WHAT LIVES INSIDE A PCB

While it may look like an unassuming piece of green plastic, a lot goes on inside a printed circuit board. Check it out.

CORE

At a PCB's center is a sheet of fiberglass-filled epoxy that has a layer of copper bonded to one or both sides.

INNER LAYER

These copper circuit traces buried inside the finished board are patterned by a process called *photoetching* before the outer layers are added.

PREPREG

This fiberglass cloth is preimpregnated with an adhesive that's activated by heat and pressure. This layer both glues the PCB's layers together and insulates them from each other electrically.

OUTER LAYER

This copper foil is applied over the prepreg then patterned by photoetching to form additional circuit layers.

VIA

Also called *plated through-holes*, these openings of various sizes are drilled after the PCB layers are laminated together. A special process applies metal plating on the inside of each hole and in a small circle around the rim on both sides. This metal offers an electrical connection between circuits on different layers, allowing components mounted on one side of the board to connect with circuits on the other, or with circuits buried in the inner layers.

SOLDER MASK

This coating, which is often clear, protects the PCB's outermost copper layers from damage caused by exposure to the atmosphere, water, or harsh chemicals used in manufacturing.

COMPONENTS

Today, most devices use surface-mount components, but sometimes old-style through-hole technology comes into play for heavy components that need a strong mechanical connection to the board.

145 FIND FACTS ON DATASHEETS

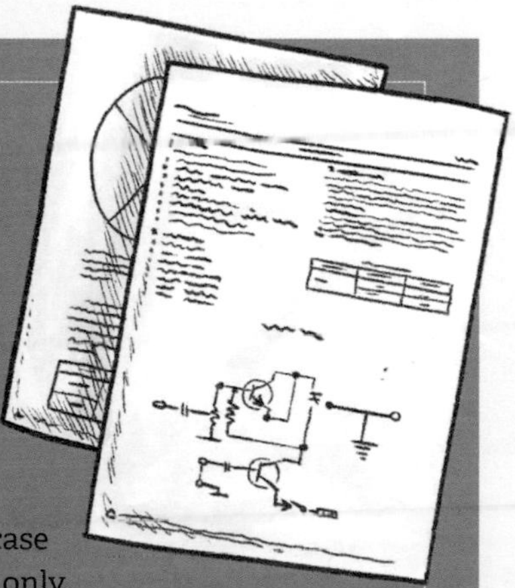

When you're designing electronics, it's crucial to choose components very carefully. Apart from obvious factors like cost and size, there are the various electrical parameters, pin assignments, and environmental conditions to consider, such as how hot you can get components without ruining them or how cold you can get them and still expect them to operate correctly. All this information is carefully tested by the manufacturer and published in a document called the *datasheet*. In the case of a simple passive component such as a capacitor, the datasheet may be only one page; in the case of a complex active component like an integrated circuit, it may be several hundred. If you've got a question about a part, you can almost always find its datasheets on the manufacturer's website or simply by Googling the part number and the word *datasheet*.

146 DESIGN YOUR OWN PRINTED CIRCUIT BOARD

The first step in PCB fabrication is drawing your circuit in an *electrical computer-aided design* (ECAD) *program*. A number of excellent ECAD programs (many available for free) allow you to render your circuit as a schematic, selecting real-world parts from a library. As you make the schematic, the software automatically lays out a circuit board, letting you switch views and updating both as you make changes. Here we use ECAD to design a version of our blue LED circuit (remember #049?) for professional PCB fabrication. You need a computer, a soldering iron and solder, a 9V battery, a battery clip, a switch, a blue LED, a 330Ω resistor, and a 1000 µF electrolytic capacitor.

STEP 1 Download Fritzing. Designed for electronic newbies, Fritzing offers schematic, breadboard, and PCB views, which means you can design all three views simultaneously. Download it for free at Fritzing.org and follow the online instructions to install.

STEP 2 Open a new document and pick the breadboard view. Click into the Parts tab and search "LED." Select any generic 5mm LED and place it onto the breadboard according to your original circuit design. Repeat with the other components, then add the jumper wires. Test the circuit on a real breadboard. Once you've got it right onscreen, test with a real breadboard and components to verify that it works.

STEP 3 Click to the schematic view and you'll see a rat's nest of part symbols and dotted lines—the software is smart enough to connect the right ends of the components but not smart enough to arrange them in a pretty way. Click, drag, and rotate the symbols until the drawing looks like the one shown here. Remember four basic rules for clear schematics:

- Use straight lines and right angles.
- Put positive voltages at the top.
- Route signals from left to right.
- Group components together by function.

STEP 4 Now click over to the PCB view, where you'll find another rat's nest. Click, drag, and rotate the component footprints until they look something like figure 4. Keep in mind that the bodies of the parts are likely bigger than their footprints, so you've got to visualize whole components as you work and make

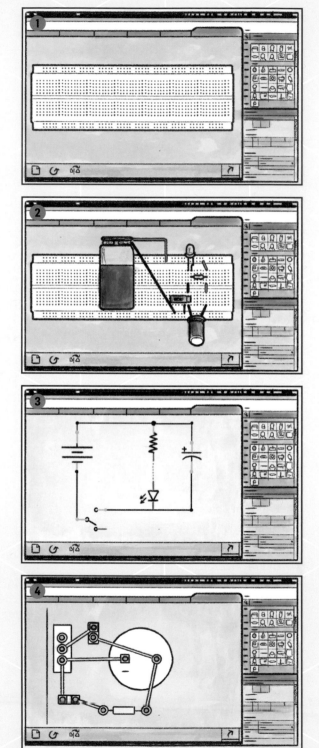

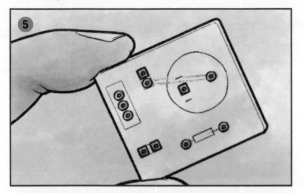

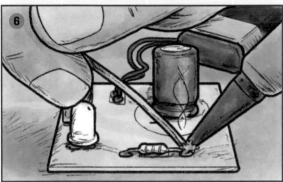

sure you leave enough room. Next, minimize the number of times the traces cross each other without connecting. When yours looks like figure 4, click the Autoroute button and watch what happens. Then select Routing > Design Rules Check (DRC). When everything checks out, save the file.

STEP 5 Order your board. The Fritzing Project runs a low-cost PCB fabrication service that makes boards from your Fritzing files—just click the Fabricate button in the lower-right corner of the layout window. (You can also export *Gerber-format files*, the industry-standard file format for PCB fabrication, for manufacturing by a third party of your choice. Know that different services have different capabilities, and you'll need to do a careful check to make sure your PCB doesn't violate the manufacturer's design rules.)

STEP 6 Once you've received your board, solder on your components and test it out.

147 WIELD A SOLDERING IRON LIKE A PRO

The connections between components and PCB must be mechanically and electrically strong. Enter the *soldering iron*, a hand tool with a metal tip that melts *solder*, a metal alloy that fuses workpieces together.

CLEAR THE AIR Always work in a well-ventilated area or with a fume-extractor fan nearby.

SPLURGE A LITTLE Spend a bit more money and get a decent digital, temperature-controlled soldering station. Using cheap underpowered irons is the number-one cause of frustration among beginning solderers.

KEEP IT CLEAN Swap out the wet sponge that likely came with your soldering iron for a dry brass wool sponge. Clean the tip after every joint.

PICK THE RIGHT SOLDER Use no-clean, lead/tin solder in the skinniest gauge you can find.

RACK 'EM UP Position your parts securely before soldering. Use helping hands or another jig, and, if possible, bend or twist the leads to form a strong mechanical joint before you pick up your iron.

FOLLOW THE ORDER OF OPERATIONS First, let the iron warm up. Heat the joint between the components, then apply the solder to the joint, and hold it until the solder flows free and bright. Don't melt solder on the iron and then smear it on.

TAKE BABY STEPS Develop some skill soldering through-hole parts before trying your hand at surface-mount components.

148
MEET SAMANTHA ROSE
INVENTOR OF THE GIR SILICONE SPATULA

For Samantha Rose, her big "a-ha!" moment as an inventor came disguised as a pesky but fairly common occurrence for the home cook: She broke a spatula while making dinner. The silicone blade—aka the business end—had held up nicely, but the wooden handle had been burned a few times and was now splintering from numerous dishwasher runs. "I looked at it and thought, 'Why don't they just make the whole thing out of the silicone that the blade is made of?' she remembers. "It would be so much better."

Rose did a quick Google search to see if she could buy what she envisioned, and was surprised to find zero results. So the self-described science nerd took the plunge: "I figured the best way to figure it out was just to try it." This entailed sculpting a model of what she thought a one-piece spatula should look like, then casting a mold with a DIY silicone modeling kit ordered online. And lo and behold, her first spatula turned out great.

So great, in fact, that Rose wanted to give some to friends and family, only made out of a food-safe silicone. She called around to a few factories, asking for a production run of ten spatulas. "It's kind of a joke and legend now, but everybody laughed and said no—you can make 10,000. That was when it went from science experiment to entrepreneurial experiment."

This lesson in minimum order quantity was just the first of many manufacturing discoveries. The second challenge was getting the material just right. "In the beginning, I thought silicone was just silicone," she says. "Then I did a lot of research into thermal and heat resistance, and I ended up talking to a rocket scientist—literally a rocket scientist—who had spent his career studying silicone for the O-rings used to seal rockets." Rose learned that competitors who claimed a heat resistance of up to 800˚F (425˚C) were just

continued on next page

blowing steam, and that a heat-proof rating of 464°F (240°C) was more realistic for a food-safe silicone product. She also wanted to exceed expectations when it came to consumer safety: "It's not like cookware companies get together and share secrets at a yearly spatula summit, but I'm not aware of any others that are using pharmaceutical-grade silicone like we are. If it's safe enough to be in your body, it's safe enough to be in your kitchen drawer."

The inside of the spatula presented some intriguing challenges as well. As it turned out, a unibody, all-silicone design would flop around unless equipped with a rigid core, which doesn't do at all when you're trying to, say, scrape down the sides of a mixing bowl. Metal wouldn't do the trick, as it would heat the tool and eventually cause it to degrade from the inside out. And plastic wasn't an option, since it melts under the high heat that silicone so famously resists. The solution? A plastic matrix, which strengthened the spatula's long handle, covered in fiberglass to withstand high temperatures. "The inside became just as important as the outside, which I never anticipated," she recalls.

While a brilliant solution, Rose and her team didn't want users to ever see or touch the fiberglass core, which ushered in a whole new problem: where to manufacture the product. While options for injection molding are both cheap and abundant in the United States, their design required *compression molding*, a technique in which heated silicone is poured into a mold, closed off, and subjected to heat and pressure. This type of molding is best done in Asia, Rose found, which meant that "we had to overcome the language barrier and work with China. We also had to overcome the barrier of 'Gee, we're taking this ecofriendly, locally sourced silicone and carting it across the ocean to be molded,' which was a bit of a marketing problem for us."

Rose didn't let that slow her roll, however. "At some point, the day-to-day challenges shift from passion for the product to logistics of getting it out into the world, like quality control, shipping, and fulfillment to customers." If that crushes you, she advises, seek out a partner—and know that it gets better. "If you can cross that hump or desert of logistics, if you can find that person who wakes up in the morning to do those tasks, then you can circle back to doing the product development you love." Rose is living proof of it—she's currently extending her line to slotted spoons, lids, ladles, and more.

Q+A

Q: What's your favorite tool?

A: Tongs. I have a tong obsession, and I *just* invented the perfect pair.

Q: Are there any other inventors or makers who particularly inspire you?

A: Scott Heimendinger of Modernist Cuisine, who invented a sous-vide immersion circulator. He's also in search of the perfect clear ice cube, so he has an aquarium circulator running in his freezer to agitate the water.

Q: What do you do when you get stuck working on an invention? How do you hit refresh?

A: I start working on something else, then I come back to it. Usually my lightning moments are completely not in a workshop—like when I'm walking down the street or in Central Park.

Q: What's your earliest memory of tinkering?

A: My dad had a jigsaw in our garage when I was little and he made all our toys out of wood. I also remember building a clock with him.

Q: What's your dream invention—if time, money, and the laws of physics didn't apply?

A: I have really vivid dreams, and I've never been able to communicate them to the outside world. I'd invent a machine that would allow me to see my dreams when I was awake.

Q: Do you have any words of wisdom for aspiring inventors?

A: Find the right people and ask them the right questions in an earnest way—it paid off for me in spades. The generosity of the community—their willingness to share their wealth of knowledge with my little project—was so cool.

Dubbed GIR for "Get It Right," Samantha Rose's dream line of silicone spatula cookware took her on an adventure in the material sciences. But she didn't feel that she needed to reinvent the wheel to achieve her goal. Instead, she tweaked this kitchen essential to make it more efficient, longer-lasting, safer to cook with, and more satisfying to use—and available in a dozen fun colors, to boot.

Rose and her team devised an internal core of fiberglass and plastic, which reinforced the spatula's handle but still allowed for flexibility in the business end.

Up until GIR hit the market, spatulas had a two-part construction: the silicone paddle at the end and a handle made of wood or metal. Rose streamlined the design by making a single, all-silicone mold.

The magic's all in the materials: Rose's silicone of choice was carefully vetted to ensure a heat safety rating of up to 464°F (240°C) and an FDA-approved food safety seal of approval.

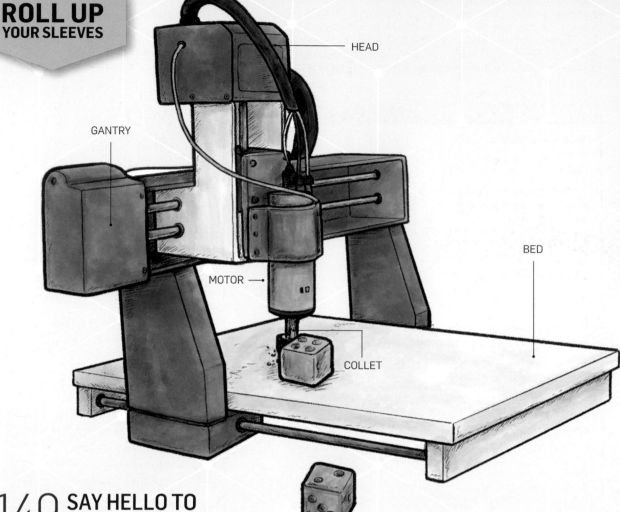

HEAD

GANTRY

BED

MOTOR

COLLET

149 SAY HELLO TO THE CNC MILL

It slices, it dices, it gouges and makes grooves! If you're looking to make rounded or sloped cuts through flat stock, make partial-depth incisions, or cut a material that emits nasty fumes under heat, a laser cutter just won't do the trick. Enter *routers* and *mills*, power tools that remove stock by rotating a sharp bit against a workpiece at high speed. The difference between the two? A router moves against a stationary workpiece (often wood), while a mill remains stationary as the workpiece—usually metal or wood—moves against it.

Today we use the blanket term *CNC* (computer numeric control) *mill* to cover any robotic power tool that uses a rotating bit to remove stock at right angles to the axis of rotation, regardless of the material and whether it (or the machine) is moving. For inventors,

these tools work wonders—you can cut gears, drill precise holes, machine custom parts, and sculpt hard stuff into organic shapes. And when it comes time to actually put your product on the factory floor, the technology you used for prototyping may not be that different from what the pros put to work.

CNC mills are often characterized by the number of axes along which you can control the cut. With a three-axis mill, you can slice in the two horizontal dimensions (X and Y) as well as vertically (Z). Pricier multiaxis mills can also rotate the orientation of the cut along at least one axis and have as many as nine controllable axes. Those big guns are fun to watch, but a three-axis mill is more than enough machine at this stage—and there's likely one at a makerspace, community college, or tool-sharing collective near you.

150 MEET PAPA, MAMA, AND BABY BEAR

CNC mills come in a wide range of sizes and capabilities. Naturally, bigger and more powerful machines command higher prices.

INDUSTRIAL/PRODUCTION

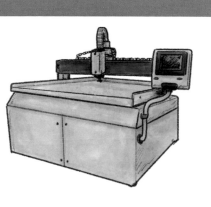

WORKING SPACE
20 by 6 by 1¼ feet
(600 by 200 by 40 cm)

APPLICATIONS Equipment manufacture, architectural work

MATERIALS Steel, aluminum, plastic, wood, composites, foam

PRICE US$20,000–$150,000

SHOP/PROTOTYPING

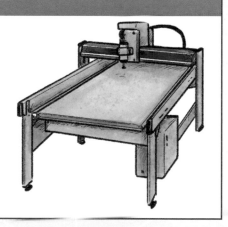

WORKING SPACE
7¾ by 4 by ½ feet
(240 by 120 by 15 cm)

APPLICATIONS Cabinetry, sign making, tooling

MATERIALS Aluminum, plastic, wood, composites, foam

PRICE US$5,000–$20,000

HOME/DESKTOP

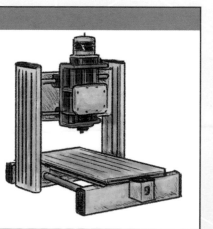

WORKING SPACE
5½ by 4¾ by 1½ inches
(14 by 12 by 4 cm)

APPLICATIONS PCB prototyping, small parts

MATERIALS Soft metals, plastic, wood, composites, foam

PRICE US$500–$5,000

151 PLAY IT SAFE WHEN CNC MILLING

With great power comes great responsibility, so read the instructions for your CNC mill before getting started, and be sure to follow these safety guidelines at all times.

COVER UP Wear hearing and eye protection gear that's up to safety regulations whenever the machine is running.

SEE RED In case of an emergency, hit the big red stop button. (Never use a machine that has no emergency stop—it's like driving a Mack Truck without brakes.)

KEEP YOUR DISTANCE Do not get any part of your body within 6 inches (15 cm) of the bit.

CHECK THE STOCK Make sure the wood, metal, or other material is free of potential projectiles, like metal fasteners.

MIND THE MACHINE Do not leave a running machine unattended.

TIDY UP Clean the machine and surrounding area scrupulously after each use.

152 MILL YOUR PRINTED CIRCUIT BOARD

If you've never operated a robot cutting tool, milling a prototype PCB is a great beginner project. Here, the CNC mill makes shallow cuts through the copper layer to carve out the circuit traces, and also drills through-holes for mounting components. You'll need access to a desktop three-axis CNC mill, a 1/32-inch (0.8-mm) end mill bit, CAM/client software for your mill, a computer, and a metal spatula, as well as a double-sided FR-1 copper clad board and double-sided tape.

STEP 1 Make the CAD model. Earlier in this chapter (see #146), we used Fritzing to lay out a circuit board for fabrication. If you haven't done that project yet, complete it through step 4 before proceeding. Once you've got the layout ready, choose the Export option in Fritzing and select the Gerber file format—this is the industry-standard file type for PCB CAD data.

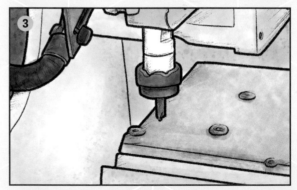

STEP 2 Run the CAM processor. Now that we've got a CAD model in a standard file format, we need to turn it into a set of instructions for the CNC robot. This step is referred to as *computer-aided manufacture* (CAM). Basically, the CAD file encodes the part's 3D shape, while the CAM file encodes the specific movements your machine needs to perform to replicate that shape. Refer to your mill's instructions for choosing, setting up, and running a particular CAM processor program—you'll need to give it information about the size and shape of the bit, the dimensions of the stock, and the locations of any fixtures attached to the bed. Depending on the make and model of your mill, the CAM program may or may not be integrated into the client software (the program you actually use to control the machine while it's running).

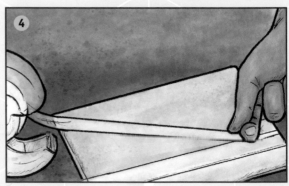

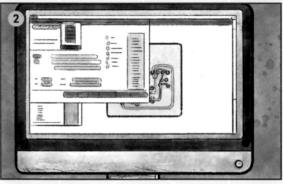

STEP 3 Set up the tool. Milling a PCB requires no deep through-cutting, so you don't need to worry about a *spoilboard* (a disposable work surface that protects the mill's bed). Since this is a double-sided PCB, you'll need to run the mill once to cut the top side, then flip the board over and run it again to cut the bottom. You must get the board in the same place for both cuts; your mill should have a fixture for accurate repeat positioning, so configure your client software to know where the fixture is. To configure, install the bit in the collet, then run the alignment routine in the client software so it knows exactly where the end of the bit is.

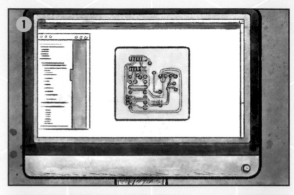

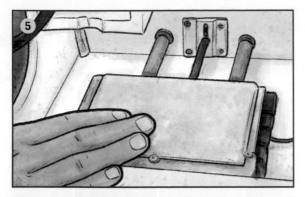

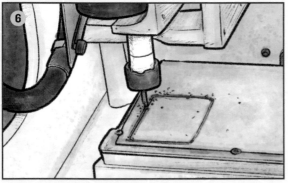

STEP 4 Clean the board, fixture, and bed carefully to remove any dust or other debris, and cover one side of the board with an even layer of double-sided tape.

STEP 5 Adhere the object to be milled to the bed against the fixture. Configure the client software to cut the top-side traces and drill the holes, but leave cutting the outline for the bottom-side cut.

STEP 6 Make the part. Once everything is set up, start the cutting operation using your mill's client software. For this small, simple board, the cut will probably take no more than ten minutes. If you've never seen a CNC tool run before, you'll want to watch it work. Whatever you do, don't wander too far off in case something goes very wrong, in which case you should first try stopping the cut from the software interface. If that doesn't work, hit the mill's red emergency stop button.

STEP 7 Once the top side is cut, pry the board off the bed with your spatula, apply a layer of tape to the top side, flip it over left to right, and stick it to the bed against the fixture again. Now run the bottom-side cutting operation. When that's done, pry the board off the bed, clean off all the tape, and set aside the scrap for use in another project. You're now ready to solder the components to your board, clip on a battery, flip the switch, and see the LED light up.

153 BUY OR BORROW A PCB MILL

As with a 3D printer or laser cutter (see #068 and #041, respectively), a desktop CNC mill is more of a luxury than a necessity. Unless you're on a very tight timetable, you'll usually find it's more efficient to send off for a custom professional PCB than to try to mill your own. On the other hand, the machine can prototype stuff other than PCBs, and is great for machining small precision parts in wood, plastic, brass, and even aluminum.

154 GET MANUFACTURING TIPS FROM PROS WHO KNOW

We've covered a lot of ground since the Industrial Revolution. Here are some kernels of advice gleaned along the way.

"I am a woman who came from the cotton fields of the South. From there I was promoted to the washtub. From there I was promoted to the cook kitchen. And from there I promoted myself into the business of manufacturing hair goods and preparations. . . . I have built my own factory on my own ground."
– *Madame C.J. Walker, creator of a successful line of beauty and hair products for black women*

ON TURNING A MANUFACTURING PROBLEM INTO A BUSINESS OPPORTUNITY: "I WAS RUNNING AN ASSEMBLY LINE DESIGNED TO BUILD MEMORY CHIPS. I SAW THE MICROPROCESSOR AS A BLOODY NUISANCE." – *Andy Grove, cofounder and former CEO of Intel and pioneer in microchips*

"We have been manufacturing overseas for two years. Recently, we've tried to update various components of our product and have struggled to iterate. This is partially due to a language barrier. But more detrimental is our inability to put our designer and our manufacturer in the same room. Overseas manufacturing may cost less, but it hinders your ability to collaborate." – Aaron Schwartz, cofounder of Modify Watches

"TIME WASTE DIFFERS FROM MATERIAL WASTE IN THAT THERE CAN BE NO SALVAGE." – *Henry Ford*

"IN THE CASE OF APPLE, THEY DID ORIGINALLY DO PRODUCTION INTERNALLY, BUT THEN ALONG CAME UNBELIEVABLY GOOD OUTSOURCED MANUFACTURING FROM COMPANIES LIKE FOXCONN. WE DON'T HAVE THAT IN THE ROCKET BUSINESS. THERE'S NO FOXCONN IN THE ROCKET BUSINESS." – Elon Musk

"One of the big failures for the big auto companies is that even the CEO and the top management often don't understand design and manufacturing. As a CEO, you have to make decisions; you need to have knowledge." – *Henrik Fisker, designer of iconic luxury cars*

"When you go to war, you need to have both toilet paper and bullets at the right place at the right time. In other words, you must win through superior logistics." – Tom Peters, business management expert and coauthor of *In Search of Excellence*

**"YOUR ONCE GRATEFUL VENDOR MIGHT IGNORE YOU WHEN HE'S GOT A BIGGER CUSTOMER, OR YOUR TIMELY GUY MIGHT HAVE A FACTORY FIRE AND GO OUT OF BUSINESS FOR A FEW MONTHS ... THE KEY IS HAVING A HANDFUL OF GOOD, OR OK, VENDORS THAT YOU CAN BOUNCE BETWEEN." –*Mary Apple, founder of Pretty Pushers, a company that creates labor gowns for women*

"IMPROVEMENT USUALLY MEANS DOING SOMETHING THAT WE HAVE NEVER DONE BEFORE."

–*Shigeo Shingo, Toyota production guru*

On manufacturing Nest, the smart thermostat, to exceed industry standards: "It's not just about turning up or down the heat. It's about the other experiences that come with turning up or down the heat—what are we doing about energy, what are we doing about your health and safety." – Tony Fadell, inventor of Nest

WORKING WITH A MANUFACTURER

Despite the impression that factory work is done automatically by futuristic robots, there are real humans at the helm—and since they have a big job, it's statistically certain that mistakes will be made. The important thing is that you work with them to build a trusting relationship so that you can fix those problems efficiently and cheaply for all.

155 INSIST ON A PILOT BUILD

A *pilot build* is a very short run of your product (perhaps even less than 1 percent of your first order's entire volume) that's used to test the waters and make sure the goods coming out are up to spec. It may seem like a no-brainer, but you'd be amazed how often small-contract manufacturing clients get pressured—either by the factory or by market deadlines—into skipping this simple step and ordering thousands of units before seeing with their own eyes that the factory's process is working right.

The manufacturing contractor loses money for every second that its machines sit idle, so the company may push back against the idea of stopping the line for you to check their work, especially if you're not a big fish. Stick to your guns. You don't want to personally open up 10,000 boxes and insert a missing part, which is really a best-case scenario for a major manufacturing error. Without a pilot build, you can very easily end up stuck with an entire inventory of product that is completely unusable—and therefore unsalable.

156 BE SMART ABOUT QUALITY CONTROL

Now hear this: The act of evaluating all products as they come off the production line (known as *quality control* [QC]) should not be an afterthought. Rather than separating this process out from assembly and packaging, treat it as an integral step from the start. As a first-time inventor, you're likely small fry to your factory partner—they'll be glad to have your business but may not stand to lose much if you don't come back. So you may need to take on more responsibility for quality control than a bigger customer.

COMMUNICATE COHERENTLY AND OFTEN Make sure both parties are crystal clear about who's doing what in terms of quality control before signing a contract or issuing a production order.

BE THE BOSS Put in the contract that the factory can't change processes, subcontractors, suppliers, materials, designs, and so on without your say-so.

CONDUCT YOUR OWN INSPECTIONS As often as you can afford to, buy and use your product just like a regular customer, keeping your eyes peeled for problems.

FUNNEL FEEDBACK Marketing, customer service, returns, and your own QC testing should all be working together with the factory to maintain a system of *quality assurance* (QA), which is the overall process of troubleshooting problems. If quality control is the safety net that catches gaffes, quality assurance is the established team effort that fixes them.

157 DROP A BOM (BILL OF MATERIALS)

Consider the *bill of materials* (BOM) a recipe for making your product. The ultimate function of the BOM is to specify—down to the smallest detail—exactly what you want mass-produced. It's the document of ultimate authority about how your product is made, and many will refer to it (both inside and outside your company) to answer the many questions that will certainly pop up during the process of bringing the factory online. So it's vital that the BOM be clear and correct.

Though several more powerful (and expensive) software options exist, a spreadsheet program is the de facto standard for BOM management and is good enough your first time out. Each assembly, subassembly, and component goes on its own line, with columns indicating all the critical details for each part. At a minimum, these columns should include an internal part number, a descriptive part name, a revision code, the quantity required and its unit of measurement, whether the part is *off-the-shelf* (OTS) or *made-to-spec* (MTS), and any special notes needed for clarity. Electronic components should also list the labels used to indicate them on circuit diagrams and PCB layouts.

And finally, your extreme left-hand column should designate a "BOM level" that reflects the hierarchical arrangement of assemblies, subassemblies, sub-subassemblies, and so on. Conventionally, the complete finished product, as the customer receives it, is the only item at BOM level 0. The product, the packaging, the instructions, and any accessories that go in the box with it are BOM level 1. The major assemblies that go together to make the product are BOM level 2. The subassemblies that go into those assemblies are BOM level 3, and so forth, until you get down to parts the factory doesn't have to put together.

Be sure to include diagrams and schematics so factory workers can assemble your product. Double-check these plans against your BOM to ensure you have the correct parts and quantities.

158 RUN A FINAL DESIGN REVIEW

Selling a retail consumer product isn't like making small batches or one-offs for a few clients you know personally. You take on certain legal responsibilities and risks when your end customer is a stranger, which you should take into account when defining quality standards.

IS IT DANGEROUS? Your first concern is civil liability. Your invention must be safe to use—yes, even for morons who may be inclined to misuse or abuse it. If the product is involved in an accident, you can be sued, regardless of whether it's reasonable or just. You'll then have to show in court that you took every step to make your product safe and to fully inform the customer of any remaining risks associated with its use.

DOES IT NEED THE LAW'S BLESSING? In many countries, there are a number of government agencies that regulate consumer products. And while there are very few goods that are completely illegal to sell, there can be serious legal consequences for misrepresenting a product as, say, a medical device when it hasn't been approved by the FDA or other relevant agency.

DOES IT NEED INDUSTRY VETTING? In less extreme cases, you may find that retailers won't stock your product unless it's been approved by some government or industry authority. Examples of markets where you need to pay special attention include products for children 12 and under, medical inventions, food and groceries, home appliances, safety equipment, electronic devices (especially those that send or receive radio signals), vehicles and parts for vehicles, power tools, and sporting gear.

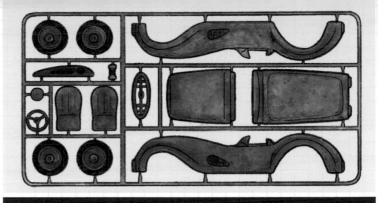

159 PAY A VISIT TO THE PLANT

In the ideal situation, your factory is literally just up the road so you can pop in to make friends, help solve problems, and learn about the nitty-gritty details of how your product goes together. And though this aspect often gets lost in the shuffle, all manufacturers and their clients have a moral responsibility for the health, safety, and dignity of the people who actually work on the line (see #161)—visiting the factory can help you sleep easier.

Still, it's not always practical to find a local manufacturing partner, and the cost of visiting yours may be inconvenient or even prohibitive. You should try to do it at least twice, anyway—once with plenty of advance warning during the pilot build and once on short notice during the production run. Being on-site during the pilot build will make both you and your manufacturer happy because having you there will make it easier and quicker to solve problems and get the line rolling again.

And a brief, friendly surprise visit during production will ensure things are still being done the right way while your back is turned. It'll also show your manufacturer that you're serious about quality and have the resources and commitment to become a longer-term customer. A polite cover story—"I'm in town for a conference," "I'm vacationing and wanted to show my family the plant," "I missed my connection and will be in town longer than I thought"—can help make this less awkward. During your visit, keep your eyes open for employees who aren't wearing proper safety gear, seem to be sleeping on-site, or look like they might be too young to vote.

160 GET THE BACKSTORY ON INJECTION MOLDING

One night in 1967, in a rented backyard in Beverly Hills, actor Walter Brooke, playing a 50-something neighbor credited only as "Mr. McGuire," put his arm around a young Dustin Hoffman, playing 21-year-old antihero Benjamin Braddock, and counseled him with just one word: "Plastics."

A few months later, *The Graduate* opened to widespread acclaim, briefly enjoying the third-biggest box office ever. For many young baby boomers, the film summed up everything that had gone wrong with the WASP-y America they'd inherited, and "plastics" became a countercultural watchword.

As director Mike Nichols has since pointed out, however, the line was already a 20-year-old joke in some circles at the time of filming. In fact, the first injection-molding of plastics happened much earlier than that, dating at least to the 1872 patent, by brothers John and Isaiah Hyatt, of a machine consisting of a heated barrel, a piston, and a mold, which they used to cast buttons and other haberdashery in *celluloid*, an early plant-derived plastic.

The first fully artificial commercial plastic, Bakelite, arrived in 1907 and was of a type called *thermosetting*, in which separate chemicals react in the mold to form a tough solid that cannot be melted again. Two-part epoxy is another example that's still in common use today.

Modern *thermoplastics* (made from one chemical that can be remelted repeatedly) were not widely successful until RCA introduced the vinyl phonograph record in 1931, and it was not until 1947—exactly 20 years before *The Graduate*—that the modern plastics industry really began to take off. In that year, James Hendry patented the first injection-molding machine with a screw in the melt chamber, which gave a groundbreaking improvement in quality control. Hendry-style screw extruders are still standard today, and he himself would go on, some 25 years later, to pioneer a gas-assisted process that would revolutionize the production of large, high-quality injection-molded parts.

161 AVOID NIKE'S ETHICAL MANUFACTURING WOES

Nike has been around since the 1960s, but the company made its first big splash in the sports and streetwear markets with the release of the Air Jordan in 1985. The celebrity endorsement of star NBA player Michael Jordan brought the brand record-breaking success in sales.

To handle the increase in orders, Nike moved its production to Asian factories, which offered faster, larger, and cheaper production runs. These factories kept their deals by paying low wages, often to children who were forced to work most days a week, often for 16 hours a day, in unsafe working conditions. The contractors hired to oversee the factories weren't based in Asia, so communication about the well-being of their workers was out of sight, out of mind.

In the 1990s, anti-sweatshop activists began protesting Nike's labor practices. Ultimately, these demonstrations made the brand name nearly "synonymous with slave wages, forced overtime, and arbitrary abuse," Nike cofounder Phil Knight lamented in 1998, the year that the company's earnings plummeted by 69 percent.

The teachable moment? The public will judge you for the practices of your contractors and business partners, so you'd better vet them for compliance with labor and environmental regulations and monitor them throughout your relationship. You can't assume the position of "Hey, we don't own the factories; we don't control what goes on there," as former Nike director Todd McKean said of early policies.

As for Nike, they turned it around, creating the Fair Labor Association to establish a code of conduct, raise pay, decrease hours, improve working conditions, and maintain monitoring.

162 SEEK CERTIFICATIONS

Plan on selling your invention outside the United States? You're gonna need to slap a *CE mark* on it. An abbreviation for Conformité Européene, this graphic indicates that your product is compliant with the European Union's health, safety, and environmental regulations, and allows the product to circulate in about 28 countries. While spot-checking is the main enforcement method, many importers won't risk the legal ramifications—not to mention profit losses—of importing a product that doesn't bear the CE mark. If you even dream of selling in Europe, work out your CE status before the first manufacturing run.

If you can identify your invention as a minimal risk product, you may be able to self-certify. In this case, your manufacturer prepares a declaration of conformity, and then they apply the CE mark in the factory themselves. If you've invented something that falls into the greater risk category, however, you and your manufacturer must seek out an agent to assess and test your product in order to receive certification. Navigating this landscape can be a real nail-biter, fraught with legalese and broken links on European government websites, so talk to someone with experience to save yourself some stress and potential delays in production.

The cute government logos don't end with the CE mark, however. If you've invented an electronic device, you'll need to conform with the guidelines of the Federal Communications Commission (FCC), which regulates electromagnetic interference. You can run your own tests during product development to make sure your radio range, power consumption, and other considerations are up to snuff, but it may be easier to contract it out. There's also *ENERGY STAR*, the internationally recognized efficiency standard. If you choose to pursue this certification, your product will need to undergo testing in a lab that's been certified by the Environmental Protection Agency (EPA).

What if your invention is a drug, chemical, medical device, cosmetic, or food additive? You're in luck: You get to deal with the U.S. Food and Drug Administration (FDA) or the European Medicines Agency (EMA). This process typically involves submitting test data from your manufacturer, but the agency may conduct its own tests as well.

163 PREVENT HIGHWAY ROBBERY

From Ninja Tortoise action figures to Poly Station video game consoles, from Mountain Frost soda to Blueberry mobile devices, the world is awash with cheap knockoffs. And while they may be funny to the casual dollar-store browser, they're a nightmare for people who have spent time, money, and energy coming up with an invention. You can make that nightmare a double if the knockoff is of such high quality that consumers won't be able to tell the difference—almost as if, say, the original manufacturer made it.

This really happens, unfortunately. Sometimes a knockoff hits shelves mere days after the real thing arrives in stores. To prevent such disappointments, avoid sending around requests for quotations to more than a few vendors. Always vet your potential manufacturing partners carefully before you give them access to your design files or physical prototypes.

You will also need contracts with partners to protect your IP from getting copied by anyone who happens to stroll across the factory floor. While an NDA (see #223) is standard practice, strengthen it with a *non-use clause*, which will prevent partners from manufacturing your product (or a knockoff of it) for anyone but you. You'll also need a *non-circumvention clause* to discourage your manufacturers from selling your product directly to end consumers. Make sure to specify monetary damages to be awarded in the event of a breach—this makes it easier for international courts and lets your partners know you mean business.

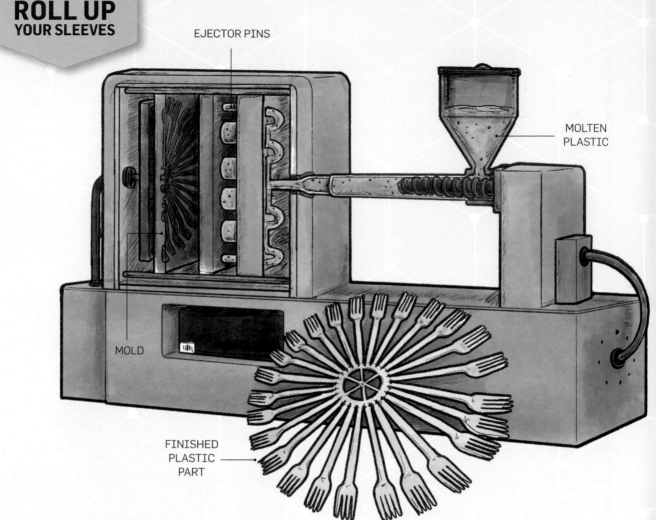

EJECTOR PINS

MOLTEN PLASTIC

MOLD

FINISHED PLASTIC PART

ROLL UP
YOUR SLEEVES

164 MEET THE INJECTION-MOLDING MACHINE

While there are an amazing variety of tools and processes available for manufacturing products today, plastic injection molding is probably the most iconic and versatile (see #160 for its history). Sure, making the mold is usually expensive, but once the tooling is ready, your per-part cost is very low. Traditionally, the setup price has limited injection molding to very-high-volume production, but the barrier is starting to creep down so that small-run operations can enjoy it too.

Broken down to the basics, there are two main phases in the injection molding process: *injection* and *ejection*. In the injection phase, molten plastic is forced into the closed mold (aka *cavity insert*), where it's allowed to cool and harden into the desired shape.

In the ejection phase, the mold is opened up and a number of pins protrude to push the solidified part clear of the apparatus. Then the mold closes again and the cycle repeats, usually once every 45 seconds or so. Often several parts are produced in one shot.

Prototyping for injection molding can be challenging. 3D printing and thermoset casting (see #066–070 and #167, respectively) are great alternatives for prototyping your invention, but when it's time to manufacture, injection molding will likely be the only practical choice for some parts. And if you haven't been planning for it from the beginning, you may be in for serious sticker shock. The sooner you wrap your head around this stuff, the better. Let's do it.

165
MOLD IT RIGHT

For simple parts such as enclosures for circuit boards, you may be able to design the mold yourself with help from your injection-molding contractor. Otherwise, turn to a plastics engineer.

PLAN FOR THE PROCESS
When designing, consider where the molten plastic will enter the mold cavity, how the air will escape from it, how the plastic will flow inside, how it'll cool, and how the solid part will be ejected.

ROUND IT OUT Avoid sharp corners inside and out, where stress can accumulate. Opt for large-radius edges instead.

STANDARDIZE THICKNESS Areas of thick plastic cool more slowly, which can cause shrinking and holes. If thickness must vary, change it gradually.

GO ON A SLANT Slope any surfaces that penetrate the mold by at least 0.5 degree so that the cooled part can eject without being damaged. This is called *draft*.

HIDE YOUR WORK Put gates, parting lines, and ejection pin pads in inconspicuous places. That way, they won't be seen in the finished assembly.

ROUGH IT UP Textured surfaces can conceal molding snafus and signs of wear.

166
CHECK OUT THE WORLD OF PLASTIC PARTS

You've probably seen all these bits and pieces before but never learned what to call them. This lineup of standard injection-molded plastic part features should help you put names to some familiar faces.

BOSSES

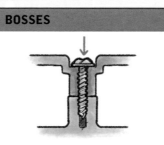

These cylindrical protrusions go around holes, usually intended to conceal, protect, and receive screws that hold parts together.

U-SNAP

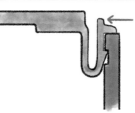

Often used to secure removable parts (like battery covers). The plastic hook flexes as it is pushed into place then snaps back into a mating catch to hold the part.

RIBS

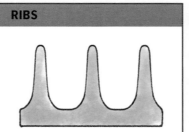

Instead of making thin walls thicker, reinforce them with ribs. Ribs should be less than five times the wall thickness in height. It's better to use more short ribs than fewer tall ones.

RING SNAP

These circular locking features secure two telescoping round parts (think of the cap on a plastic milk jug). Their flexibility makes them easy to remove without breaking.

BEAM SNAP

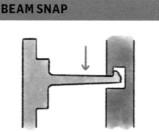

A beam snap consists of a plastic arm with a hook at its end that deflects and then snaps back into a mating catch. These snaps are sometimes designed for one-time use.

LIVING HINGE

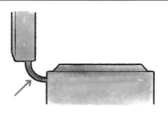

This thin strip flexes repeatedly and without breaking, providing a hinge between two pieces of a single part. They allow parts to be made in a single step and in one piece.

167 CAST A THERMOSET FACSIMILE PART

If your design has custom parts that will be manufactured by injection molding, you probably won't have access to them during prototyping. Instead, you'll need to produce facsimile parts—which have the same shape but a different composition—to stand in for those bits during testing and marketing. *Thermoset casting*, in which a two-part plastic resin sets up at low temperature in a mold made from low-cost materials, is the traditional solution.

To make your own facsimile part, you'll need the following materials: urethane RTV rubber, a ¼-inch- (6.4-mm-) thick sheet of clear acrylic, non-hardening plasteline clay, dry rice, spray-on mold release, rubber bands, disposable measuring cups, a pen cap, and polyurethane resin. You'll also need a handful of tools: a hot-glue gun, a plastic scoring tool, a straightedge, a putty knife, a hobby knife, and a permanent marker.

STEP 1 Build the mold box so it's not much bigger than the part you're casting. (This'll save you rubber.) Score and snap the acrylic sheet into five pieces: a large baseplate and four smaller side walls, each as wide as your mold will be tall. Hot-glue the four walls together as shown, and then glue them to the baseplate. Make sure all edges are completely sealed by glue so liquid won't leak out.

STEP 2 Embed the part. Press non-hardening clay into the box, filling it halfway. Make sure to tamp the clay down and completely fill the box, leaving no voids. Now press your part into the clay, leaving the part above the mold line exposed. Use a pen cap to make a series of depressions in the clay around the part. When you pour the RTV, it will fill these holes to form interlocking "keys" in the mold halves, making it easy to put them together in exact alignment.

STEP 3 Pour the top mold. First, pour rice over your embedded part until the box is full. Then dump the rice into a measuring cup and spray mold release over the embedded part and the clay. Note the volume of rice and add equal parts of the RTV rubber reactants to mix the same total volume of liquid rubber. Pour this gently into a corner of the mold box, covering the embedded part slowly to avoid trapping air bubbles against it. Let the rubber cure overnight.

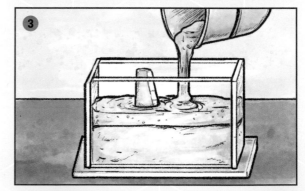

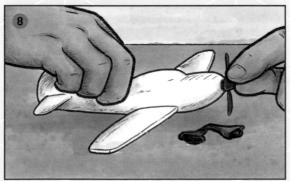

STEP 4 First, use your putty knife to break the hot-glue seal between the walls of the mold box and the baseplate. Set the baseplate aside, flip the box over, and carefully scrape out the clay. Then spray mold release on the part embedded in the cured top half, and repeat step 3 to pour the bottom half.

STEP 5 Finish the mold. Once the rubber has cured, pull the mold box apart at the glue joints, pry apart the two halves of the mold, and remove the part. Use your hobby knife to cut any vents, sprues, and runners—they will connect the empty part volume with the edge of the mold so that you can pour the part-casting liquid neatly and all air inside can escape.

STEP 6 Score and snap two pieces of acrylic sheet so that they fit exactly over the large flat faces of the mold—this is to stiffen them and make it easy to apply uniform clamping pressure. Assemble the mold, using the "keys" to align the two halves. Wrap it tightly with rubber bands to hold them together, and fill the part cavity up with water. Now dump the water out into a measuring cup, and write the volume of the interior on the mold itself in permanent marker.

STEP 7 Pour your parts. Take the mold apart, make sure all its surfaces are completely dry, then apply mold release to both halves of the part cavity. Reassemble the mold, clamp it with rubber bands again, and mix two equal parts of the polyurethane resin to give the total volume you just measured. Pour the resin in the mold, let it cure, disassemble the mold, extract the part, and cut off sprues and runners by hand. Repeat to make as many copies as you need.

STEP 8 Finish your parts. Cut off the sprues and runners, then file away any visible mold lines, excess material (known as *flash*), or other imperfections. Not every shape can be cast in a two-part mold, so you might have to drill holes, cut openings, or make other modifications by hand afterward. Or you might have to cast shapes in multiple molds, then fasten the castings together to get a complete facsimile part. At the production scale, such complex shapes can be injection-molded in a single operation using a multipart cavity insert, which is pretty neat to see.

168
MEET ASHOK GADGIL
INVENTOR OF UV WATERWORKS AND THE BERKELEY-DARFUR STOVE

As a physics graduate student at UC Berkeley in the 1970s, Dr. Ashok Gadgil had a quandary: "I was doing my thesis on the general theory of relativity. But I couldn't see any practical consequences of any knowledge I would advance in my lifetime. How would I justify the waste of education and resources invested in me by society?"

Fast-forward 40 years, and it's safe to say that society has seen ample return on its investment. Now a senior scientist at Lawrence Berkeley National Labs, Dr. Gadgil has dedicated his career to creating simple, sustainable solutions to problems of tremendous reach and scale, often benefiting poor communities and the environment. His efforts have been much lauded—in 2014, he was inducted into the National Inventor's Hall of Fame, and he's received a lifetime achievement award from the Lemelson-MIT Program. One of his first projects put 150 million high-efficiency LED lamps in low-income homes in his native India in just one year. In a more recent effort, he's used iron to combat toxic arsenic levels in drinking water, which affects 100 million people worldwide and has been deemed the largest mass poisoning in history.

For Dr. Gadgil, water contamination is a near-constant focus. One of his most celebrated inventions is UV Waterworks, a compact tabletop unit that exposes contaminated water to UV light at the rate of 4 gallons (15 L) per minute, filtering out *E. coli* and other harmful bacteria. His gadget itself is a game-changer—one that earned him an innovation award from *Popular Science* in 1996. Equally important, however, is his understanding of the economic and behavioral motivations driving both consumers and providers, who tend to be adverse to risk and new technology in these regions. Dr. Gadgil explains: "I knew it had to be a community system—it is impractical for people

continued on next page

to buy the device when they can just purchase the water." This way, a local facility could purchase the machine and sell water as a service, sparing the village people the cost of buying a device themselves. The daily price of the service? One cent per person per day for 1 L of clean water. "People who earn US$1 a day doing manual labor can buy good health for 1 cent a day," Dr. Gadgil says. "When they think of how many days of productive income they lose due to sickness, it becomes a no brainer."

Dr. Gadgil licensed UV Waterworks to a company that aims to make a profit, which assured him that they would continue to support the program. He then worked with a design for manufacturing firm to make the device cheaper and faster to build without sacrificing the science or efficiency. There were, of course, hiccups. "When I first designed UV Waterworks, my inclination was to make it difficult to access the inside so untrained people couldn't tinker with it," Dr. Gadgil recalls. "I had assumed that the heat from the electronics would prevent condensation from forming inside the device. But it turns out, when you turn the machine off, it cools down, and then water condenses on all surfaces." This caused trouble in the field. So Dr. Gadgil and his DFM team went back to the drawing board, ultimately solving the problem by piggybacking the electronics onto the device's exterior.

Ten years after he began work on his filtration project, Dr. Gadgil fielded a phone call about the situation in Darfur, where brutal civil war and genocide had driven millions of refugees into camps. A horrific trend had emerged there: Each night refugee women were being raped as they searched for cooking firewood outside the camps' walls. Seeking to again solve a humanitarian problem with an energy-efficient solution, Dr. Gadgil thought of the women's stoves: primitive three-stone setups with 5 to 8 percent efficiency ratings. "If I could push that to 20 percent," he recounts, "then maybe the women would have to go out only once a week," dramatically lessening their exposure.

He traveled to Darfur at the height of its troubles to work directly with the women, creating a design that was tailored to their needs. Then he took it one step further: "We spent a lot of time designing the stove with straight edges so that it could be precision cut in Mumbai, then shipped to Darfur where people could assemble it with simple tools. Local employment generation, local skillset building, and the stove is only US$20 bucks. Everybody wins."

Q+A

Q: What's your favorite tool?

A: Google.

Q: What's your earliest memory of tinkering?

A: When I was growing up in India, there used to be a builder's set called a Meccano, which is like a tinkerer's set where you have nuts and bolts and steel plates and all kinds of perforated things that you can build whatever you want out of. And chemistry sets too. They were a tremendous amount of fun for the longest time.

Q: Are there any inventors or makers who have really inspired your work?

A: It's a long list, but the greatest names are ones like the Wright brothers and Edison who did amazing things with what now, looking back, seems like pretty primitive equipment.

Q: What's your dream invention—if money, time, and the laws of physics weren't an issue?

A: I would want to invent a whole slew of things that allow us to live in prosperity in harmony with the environment. Because that's the biggest disaster that's coming down the pike. Unless we figure out how to live lightly on the planet, this century will be the decision point.

Q: Do you have any words of wisdom for aspiring inventors?

A: Don't be afraid to dream big. It also helped me to stretch to shake hands with people who are positioned closer to the market side and understand their language, and to speak with people who do sociological or applied economic work to understand what people choose to spend their money on.

Over the years, Dr. Gadgil's team has iterated on the Berkeley-Darfur stove to get it just right for the many Sudanese cooks who rely on it daily. Here are some of the features that have made it a staple in Darfur's refugee camps—and improved the lives of thousands.

Wooden handles do away with the need for pot holders.

Ridges in the design create space between the stove and the pot, allowing for optimum air circulation.

A tapered wind collar boosts fuel efficiency while accommodating a range of pot sizes.

Metal tabs mean a cook can also use the stove to bake bread on a plate.

The opening for the fire box was designed to be deliberately small so users can't overload the stove with wood. The air openings in the fire box are also unaligned with those of the outer stove, keeping out wind.

Users can put stakes through the wide, steady feet to prevent the stove from tipping or being blown over.

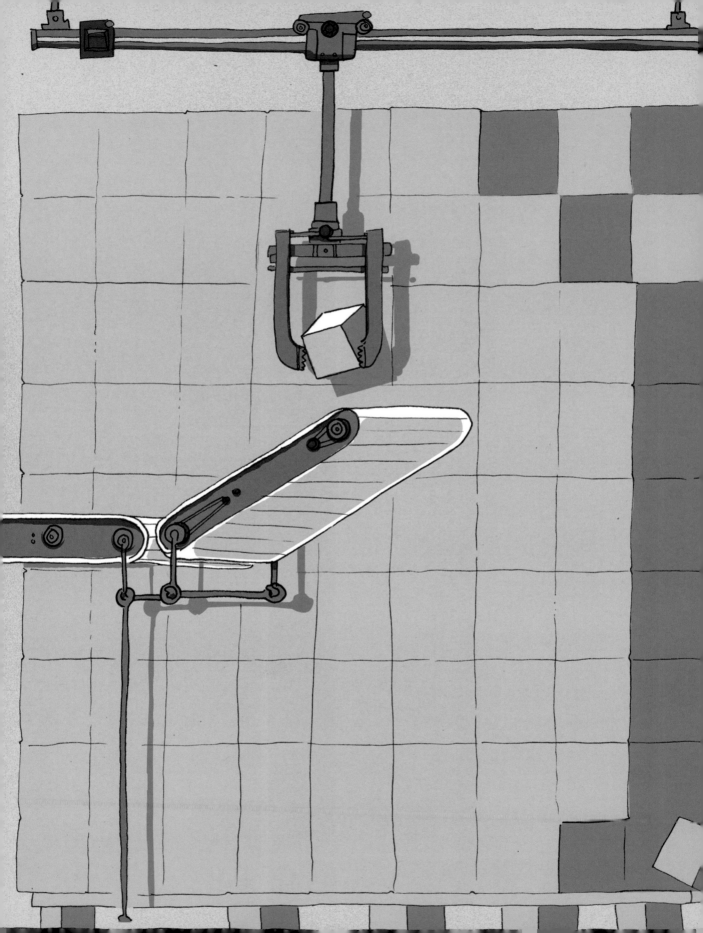

SUPPLY-CHAIN MANAGEMENT

You've thought of your invention as a prototype, a business venture, and a manufactured product. Now it's time to start thinking in even bigger terms: Your product is an economic system consisting of a network of transactions with partners both upstream (where your supplies come from) and downstream (where your demand gets fulfilled). The art and science of optimizing that system is called *supply-chain management*.

: :

169 CHOOSE PARTS WITH AN EYE TO SUPPLY

If you're absolutely sure that you're only going to make one run of your product, then yes, you can buy discount, discontinued parts in bulk, at auction, with an eye toward maximizing profit margins. But if there's any chance that you'll want to do another run without implementing a total redesign, be sure to think about more than just which part works best and costs least.

In electronics especially, it's all too common for a team to build a product, experience success with the first run, and then discover that some key component has been discontinued by the manufacturer once the team goes back to build more. This is not always an unsolvable problem, but it's certainly not ideal—it leaves you scrambling to find a replacement part that may not work as well with your design or may prove to be outrageously expensive.

So when you first draw up your production BOM (see #157), take a moment to review all your parts: who makes them, where they ship from, and how secure that supply is likely to be in the future. It's a brief effort that can save a lot of pain later.

170 OUTSOURCE, OUTSOURCE, OUTSOURCE

Decades ago, companies that manufactured products would often run their own factories and warehouses for storing parts, supplies, and ready-to-ship products. But in the 1990s, business folks began to realize that manufacturing and fulfillment were specialties of their own, and that it usually made more sense to outsource those functions to experts and instead focus on core competencies. But if your business is very successful, it may be sensible for you to consider buying or building your own factory, warehouse, or shipping center at some point down the line.

Today, however, especially for a manufacturing startup, usually the best option is to have your contract manufacturer ship completed products to a contract fulfillment center, which then ships them to retailers or customers. You may, however, want to dip into the process at various points to do independent quality control and make sure your contractors deliver what they promised (see #159). A *service-level agreement* (SLA) helps you define your contractors' duties, quality standards, and monitoring systems—and the repercussions if they fail to hold up their end of the deal.

171 PUT A BAR CODE ON IT

A *stock-keeping unit* (SKU) is a single type of thing that you sell. If you have one product, you have one SKU, which gets a unique number or other identifier to distinguish it from other products in your inventory. You can assign your SKU numbers internally, or you can set them up to correspond to a standard used by distributors and retailers, such as the *Universal Product Code* (UPC), a 12-digit unique number system used around the world.

These days, the SKU is almost always marked on every product in inventory in both human- and machine-readable (i.e., bar-coded) form. And while you should reserve the SKUs for your product inventory, there may be other valuable assets you'll want to track with unique numbers and bar-code labels—parts, tools, supplies, equipment, and so on.

Even if you're not using a computerized inventory-management system at first, bar-coding is easy and cheap. Organize a system from the very beginning, in anticipation of growth, that assumes you'll eventually be using bar-code scanners. Another option is to tag your goods with *radio-frequency identification* (RFID), tiny programmable transponders that activate when they're near a radio antenna so humans don't have to scan each item by hand.

TOOL TIME

172 COUNT PARTS WITH A COUNTING SCALE

Whether you're stocking, packaging, or shipping, a good-quality digital scale is one of the most useful tools you can keep around the factory, office, or warehouse. If you need to handle lots of small parts—such as electronic components, fasteners, or other bits of hardware—a special type of digital scale, called a *counting scale*, can be especially handy.

A counting scale includes dedicated electronics that allow you to preprogram it with the weight of a single part—whether it's a resistor, microchip, or rivet—then automatically calculate, from the weight of a pile on the scale, how many units it contains. So instead of having to manually count out 150 tiny surface-mount LEDs, you can just scoop a bunch of them onto your counting scale and add or take away a few as needed.

Weighing your finished, packaged products is also a quick and easy way to do quality control checks—if anything is missing, the package will come in underweight and should be opened and inspected to find the problem.

173 RUN A TIGHT SHIP

If you're doing your own distribution, you'll need to optimize your space to keep everything running as smoothly as possible.

PUT SAFETY FIRST A warehouse is a fast-paced environment with lots of heavy stuff zooming around. Do what you must to make sure nobody gets hurt, whether by sudden traumatic injury or chronic problems caused by bad ergonomics. As foreman or -woman of the floor, you also need to provide proper safety gear (such as hard hats, eye protection, and forklifts for lifting large loads), ensure that your sprinklers work, keep the floor clear of perils, and zone off hazardous materials. Check with your local and national labor watchdog groups to make sure your space, signage, and policies are up to code.

PLAN THE FLOOR CAREFULLY Shipping, receiving, storage, and office spaces all need to be separate and well defined. If you're tripping over people making phone calls or answering email while you're trying to pick and pack an order, you're doing it wrong. (See #179.)

DON'T FORGET RETURNS Like it or not, sometimes stuff is gonna come back. It will need to be received, examined for quality control, and restocked if practical. Plan for that process from the start.

INVEST IN GOOD EQUIPMENT This includes materials-handling gear like dollies, trucks, and lifts, as well as shelving (aka *racking*), which should be safe, tough, and designed to use space efficiently.

USE COMPUTERS You should be able to figure out instantly exactly what you've got, how much, and where it all is. Without computers, inventory-management software, and bar-code scanners, it's hard to achieve this in a bedroom closet let alone a massive warehouse.

KEEP THE DOCK CLEAR Ideally, nothing should pile up outside of designated storage areas, but in the loading dock this rule must be absolute. Nothing sits in the dock. Ever.

SCHEDULE DELIVERIES It sounds tedious, but if you don't have separate shipping and receiving docks, or you're not keeping a full-time warehouse staff yet, missed deliveries and interfering inbound and outbound trucks can cost hours—or even the whole day.

MIND THE DUST As long as you don't let it become a hazard, the natural accumulation of dust on your inventory can be a useful indicator of what you're overstocking. The thicker the dust on something, the harder you should be looking at it.

MAKE TIME FOR MAINTENANCE This means not just facilities maintenance—fixing broken lights and equipment, reorganizing to meet changing needs, and so on—but also inventory maintenance. You'll have to do a complete stock count once a year for the accountants, anyway, and that process will be a lot less painful if you develop a routine of informal rotating monthly audits.

174 MANAGE YOUR INVENTORY WITH SOFTWARE

An *inventory-management system*, or stock-keeping system, keeps track of what your business owns, including (but not limited to) finished products. Today, such systems are almost always computerized. For small operations, paper-based systems can work, but with the plummeting cost of hardware and software, it's worth your while to start with a computerized system.

Inventory-management software is useful, firstly, for knowing exactly what you've got and where it is. When transactions affect stock levels—supplies received, parts committed to manufacturing, finished products shipped—the system automatically updates the records and lets you know if it's time to reorder.

Monitoring your inventory requires keeping enough product on hand to last your *lead time*, the number of days needed to manufacture and ship more. It's also wise to have *safety stock*: extra reserves in case of a sales bump or production delay. To decide when to reorder, your software multiplies the lead time by your forecasted daily sales to predict how much product you'll ship during that period. Add some safety stock to this figure, and when your coffers dip to that sum, you'll be prompted to reorder. Just how much depends on your annual sales, order cost, and cost to "carry" the stock (i.e., storage fees, interest payments, and any taxes that apply).

175
TAKE A LESSON FROM LEGO

Started as a sideline in wooden toys by a struggling Danish carpenter in 1932, the LEGO Group is today valued at more than US$15 billion and has annual revenues exceeding US$5 billion. Since the introduction of its now-iconic interlocking plastic brick in 1958, more than 400 billion LEGO elements have been molded, and an average of seven LEGO sets are sold around the world every second. A huge part of the toy's success is its modular approach: When LEGO wants to introduce a new set, most of the parts it needs are already designed, tooled, and ready to manufacture.

The majority of the bricks are still made in LEGO's original hometown of Billund, Denmark. The factory there covers 120 acres (50 ha) and consists of three main areas: 14 huge silos for storing plastic beads, a dozen production lines with 100 injection-molding machines each, and four huge storage areas with a combined capacity of 1 billion bricks. The warehouse reaches 75 feet (23 m) tall with 42,000 acres (17,000 ha) of shelf space holding almost half a million separate bins. Sixteen robot cranes collectively patrol the warehouse, moving almost 20 billion bricks a year, each of which sits in storage an average of 2½ months before moving on to its final destination in a kid's toy bin near you.

176
PACK A PALLET

A *pallet*, aka skid, is a sturdy platform designed to move a group of single freight pieces as one unit. Designed with spacers and openings that allow them to be easily picked up by a forklift, pallets are usually made of wood or plastic, though metal pallets do exist for extreme environments or ultraheavy loads. There's no single standard size, but 40 by 48 inches (102 by 122 cm) is the most common. If you've never done it before, you should try packing at least one of your own pallets for shipment. Here are a few tips.

STEP 1 Choose it wisely. Prior to packing, make sure your pallet is undamaged and strong enough to support the load. A pallet that breaks during movement can cause serious injuries and lots of damage to equipment and freight. The pallet should also be big enough that the load doesn't overhang anywhere.

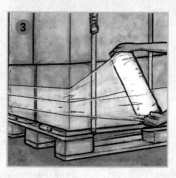

STEP 2 Load it properly. Make sure all your boxes are securely sealed and rated to support the part of the load they're going to have to bear. Put big, heavy items on the bottom and lighter and smaller stuff toward the top. Where possible, boxes should be stacked in a staggered fashion—think bricks, not plates.

STEP 3 Strap it securely. Use banding or strapping to hold the boxes on the pallet. Straps should run under the pallet top deck, not under the very bottom, and should be cinched down tight. Then wind plastic stretch wrap tightly around the entire load, lapping each turn halfway over the one before.

177 AVOID APPLE'S POWER MAC PROBLEMS

Perhaps the most crucial question in supply-chain management is how much product to manufacture and when. Storing inventory is expensive, and if you overestimate demand, you could end up with a warehouse full of product that nobody wants. In electronics, this can be especially costly because devices become obsolete fast.

In 1993, that's exactly what happened to Apple with the launch of its PowerBook line of laptops—the company overestimated demand and ended up holding a bunch of surplus inventory that cost more than it made. So in 1995, when Apple was stocking up for the preholiday launch of the new Power Macintosh (later Power Mac) series of desktop computers, the company played it safe and bought low—so low that this time, it got burned by exactly the opposite

problem. The Power Mac sold like hotcakes, and Apple's stock was quickly gone. Its suppliers weren't prepared to meet exploding demand and at one point, the company carried US$1 billion worth of unfilled orders on the books.

There was a storm of customer outcry and a wave of shareholder lawsuits, and Apple lost most of its share in the desktop PC market. CEO Michael Spindler was ousted and replaced by Gil Amelio, whose yearlong tenure saw Apple's stock sink to 1985 levels. It would take Steve Jobs's return, and a decadelong repositioning in the mobile market, to resurrect the company.

What to learn from this tech giant's mistake? For any new product, base the size of your first run on presales and the track records of similar items, and ensure in advance that your supply chain is ready to respond to increased demand.

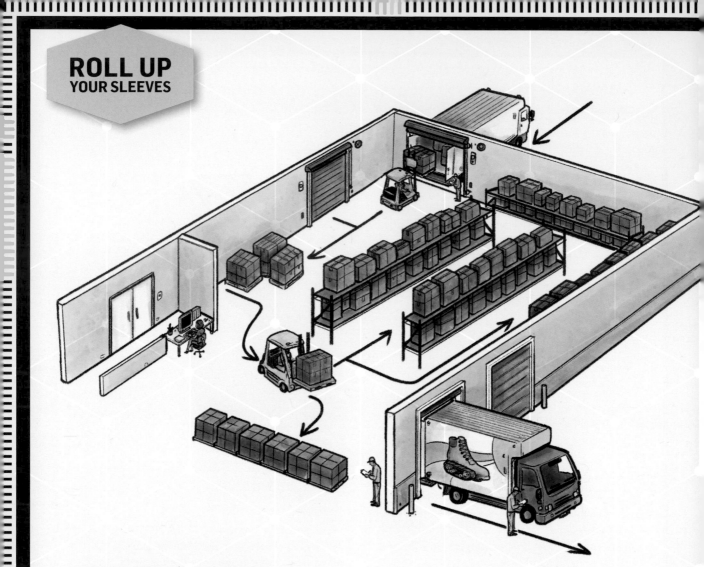

178 MANAGE YOUR WAREHOUSE SPACE

Whether you're storing parts, products, or both, running a warehouse is an expensive proposition. That's why, though each case is unique and there are exceptions to every rule, your first move should be to consider outsourcing this job to experts (see #170). If you decide to DIY, you must use every bit of space and time in the warehouse as efficiently as possible to keep costs down. Obsessive planning may seem tedious at first, but the little savings add up over time.

Warehouse operations include receiving, stocking, picking, packing, and/or manufacturing, shipping, and returns. Of these, *picking*, which is the process of going into the warehouse and coming out with what you want—whether it's parts for manufacturing or products for fulfillment—is the most costly and most

likely to reward careful planning. This will not only be important as you start but should involve ongoing efforts to improve efficiency and eliminate waste moving forward.

Many variables impact picking costs—including the information technology used to handle pick requests, the physical technology used to move stuff around, and the human resources technology used to manage employees—but warehouse layout is the most fundamental. This includes both where you put the shelves and where you place the items on them. So when you first draw up a floor plan, think carefully about what your warehouse needs to do for you and how, and make sure to consider how you'll expand operations when your business grows.

179 PICK A WAREHOUSE FLOOR PLAN

Sure, it's just aisles of shelves and tables spotted with tape guns. But where you put them matters, as it determines your efficiency. Activity in a warehouse should flow like water down a river, with parts and finished products alike floating between stations without getting snagged in the weeds by the banks.

BRANCH

Workers pick from shelves on both sides, or pick from one side and stock on the other. Manufacturing and/or packing happens at the near end of the central aisle, along with shipping and receiving.

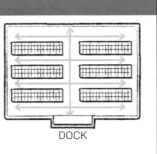

BEST USE If you only have one dock for both inbound and outbound traffic or if you can't afford a lot of special equipment.

ZONE

Also known as *ABC picking*, each stock item is classified as A (top 20 percent of picked items by volume), B (middle 30 percent), or C (bottom 50 percent) and is shelved in bulk quantities (whether as cartons or whole pallets) as well as individually.

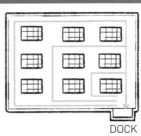

BEST USE When there are wide variations in the number of units in your pick requests.

LINE

A gravity-feed or other push-through shelving system works well because restocking and picking can happen at the same time without interference.

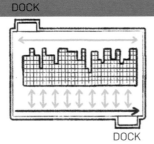

BEST USE If receiving and shipping docks are catty-corner on opposite sides of the building, especially where the value-adding operation can be arranged in an assembly or packing line.

U-FLOW

Goods flow from the receiving dock to the racking area on one side, then in the opposite direction to the manufacturing, packing, and/or shipping areas near the dock on the other side.

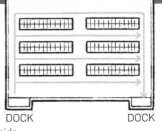

BEST USE If your docks are on the same side of the building. This layout allows for moving products straight from receiving to shipping without restocking.

SERPENTINE

The lightest, most popular items go near the front of the snake, while the heaviest and least popular items go near the end. Workers enter and exit the path as required for each request.

BEST USE When your pick requests involve a lot of different items that vary widely from one request to the next.

THROUGH-FLOW

Manufacturing, packing, and other operations occur in a straight line between separate receiving and shipping docks. The most-picked stock goes near the middle; the least popular farthest away.

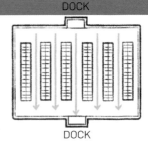

BEST USE If your shipping and receiving docks are located directly opposite each other across the building.

180 PACK A KIT

If you've never developed or manufactured products before, a kit-based business is a great place to start. Unlike ready-made goods, kits are often practical to manufacture and distribute yourself without any outsourcing, since the process of manufacturing a kit requires only procuring the parts and putting them in a box for the end customer. You may also want to kit parts for shipment to a contract manufacturer, that can do final assembly and packaging.

STEP 1 Order the parts on your bill of materials (see #157) and count them carefully when they come in. Once you've verified that you got what you paid for, enter the count into your supply-inventory system. This could be a clipboard you hang on the wall, a spreadsheet on Google Drive, or a dedicated parts-management program. Tip: If you've got lots of small parts, a counting scale is very handy (see #172). You can also use a precision digital scale and a bit of math to do the same thing: Just zero it, put ten parts on the platform, and divide the total weight by ten to get the average part weight. Then you should be able to count quite accurately by dividing the weight of any unknown quantity of parts by that number.

STEP 2 Store your parts. We'll assume the simplest case, which is one worker packing one type of kit (i.e., one SKU) from a supply dedicated to that specific purpose. Each BOM line item needs its own bin, preferably one that's wider than it is tall so it's easy to reach the bottom. If your BOM calls for multiple picks from the same bin—say your kit contains six identical LEDs—it's worth your while to pre-kit these in plastic bags so your worker can just grab one baggie and go.

STEP 3 Set up the line. Start with a table at a comfortable standing height, propping it if needed so it's just right. If one table isn't long enough, arrange two tables in a V or three in a U. Your worker should have to move around as little as possible. Put an anti-fatigue mat on the floor where she will stand.

STEP 4 If you haven't pre-kitted multiple identical parts, make sure the bin is clearly marked with the quantity that should be picked from it—put a big X with that number on the front in permanent marker or with a label maker. Now arrange the bins in packing order from left to right across the back of the table(s).

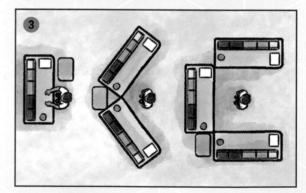

STEP 5 Set up your digital scale at the end of the line and mark it with the weight of a correctly packed kit. Set up a dedicated place for mispacked kits, and put it far away from the end of the line so there's no chance mispacks will accidentally find their way into stock.

STEP 6 Figure out how many kits you want to pack, count out your kit cartons, and stack them up at the front of the line. (Picking up and putting down a tool—like a glue gun or a tape dispenser—costs time and effort that quickly add up to waste if you have to do it every time through the line, so you might consider pre-building all your cartons. Ditto for sealing up the packed cartons at the end of the line—you might want to tape them all up at the end.) Apply your inventory label, which should have the name and SKU number of the kit in both human- and machine-readable (bar-code) format.

STEP 7 Stuff your first carton, moving from left to right as you pick through the bins. When you're done packing, close the box and weigh it. If it weighs right, put it in the finished pile. If it weighs wrong, put it in the mispacked pile. If you're getting more than one in five mispacks, stop and figure out the problem. Otherwise, crank through until you're out of boxes.

STEP 8 Tape or otherwise put the final seal on your finished kits, move them to your inventory storage area, and log them in your system as stock. Then break open your mispacks, sort their contents back into your parts bins, and update your parts inventory record based on the number of correctly packed kits you just stocked. Repeat from step 7 as necessary to make up for mispacks. Finally, review your parts inventory records and decide if it's time to reorder anything.

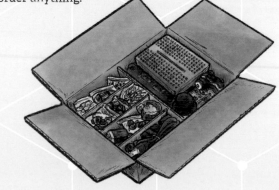

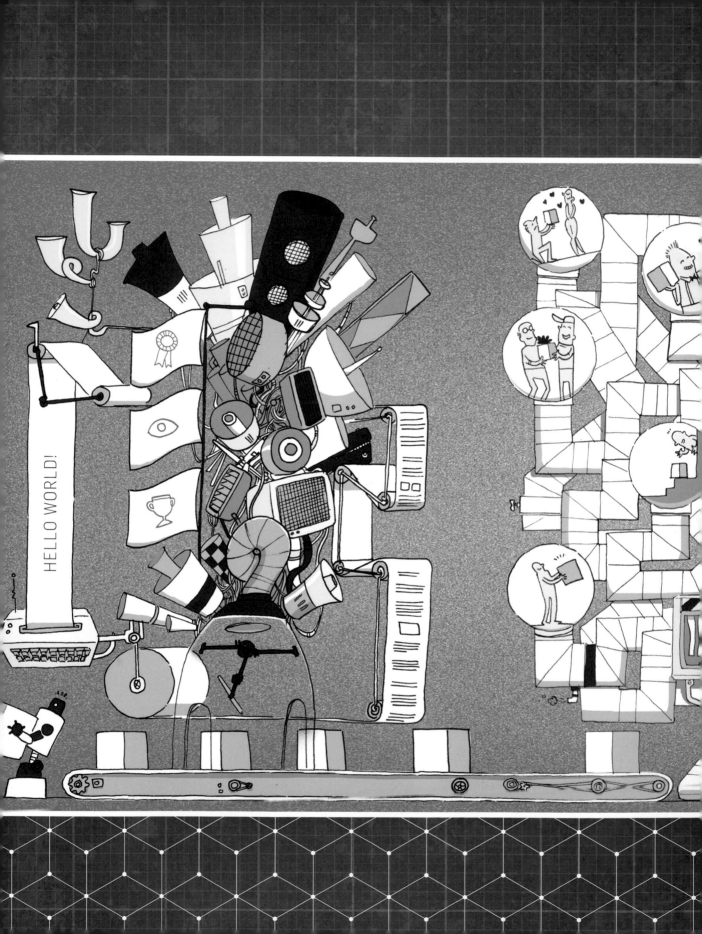

MAKE IT TO MARKET

There are many ideas about what the word *marketing* really means. One definition, dubbed "the four Ps of marketing," says it's the art and science of selling the right product, in the right place, at the right price, at the right time. (Yup, the "p" is silent in that last one.) Jokes aside, the four Ps are a great way to start thinking about this vital process.

181 SELL THE RIGHT PRODUCT

At this stage, you should have confidence that your product is a sure thing. This means not just making a worthwhile widget but understanding why it's the right product for your customers. Small businesses often can't develop deep quantitative data, but you should have done enough qualitative testing to build a convincing picture of your customers' needs (see #059). And if you've run a successful crowdfunding campaign (see #089), your backers can give valuable feedback about whether or not you're on the right track.

Once you know those things, you're ready to decide on a name, if one hasn't already come to you in lightning-bolt form. Consider what the name of your product will sound like when people say it aloud and what they'll type in to find it online. It should be unique, memorable, and evocative—not just of what the thing does but of how it's different from other products. Some choose to highlight their product's authenticity by naming it after a place or person. Other ways to distinguish your product name include inventing a new word, splicing together known words to make a compound, and tacking on a prefix or suffix, like Apple's famous iPod.

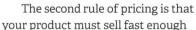

182 SET THE RIGHT PRICE

The first rule of pricing is that you must cover the costs of running your business plus any profit you hope to generate. Many small manufacturers turn to the *cost-plus pricing strategy*, in which you calculate your unit cost and multiply it by 1 + your desired markup percentage. (So if your markup is 50 percent, you'd multiply your cost by 1.5.) This model works so long as you accurately account for your costs, tallying not just the labor, materials, and shipping that go into one unit but your overhead expenses as well. Overhead includes recurring fixed costs like rent, utilities, salaries, and insurance, as well as one-time or variable expenses like advertising campaigns and repairs. Another model, *value-based pricing*, focuses less on covering costs and more on what consumers have shown they will pay.

This tactic can lead to higher profits, but it works best when there's a pre-established market.

The second rule of pricing is that your product must sell fast enough to keep up with expenses, so carefully forecast and track sales to ensure you're covering costs in the long term. If you're coming out ahead, congratulations! The difference is profit you can use to grow your business.

It's more likely, however, that you'll need to trim costs until you're at the price your market demands. It may be okay to run at break-even for a while, but a business must grow to survive, which entails lowering your costs even further while raising sales volume in order to increase and multiply profits.

183 GROK BASIC MARKETING DOS AND DON'TS

It's a big, noisy world out there, and it can be hard to know where to concentrate your efforts. Here are some tips for rising above the roar.

DO

HYPE THE LAUNCH Treat your product's on-sale date like your firstborn's first birthday. Go on a social media spree, issue a press release, and coordinate influencer and media outreach so they publish on time.

CRAFT A SELL SHEET Send retailers and marketers a one-page document that states your product's benefits in a single line, plus a nice photo and key selling points. This gives them a tool they can refer to and share.

STICK TO A SALES PLAN Forecast how many units you hope to sell each month, then choose marketing tools to target customers and sales accounts. Attach your goals to deadlines so you can make them happen. To make sure your marketing dollars are well spent, monitor sales closely.

ATTEND TRADE SHOWS Here's where you meet the big kids on the block. Whether you're looking for another round of investment or trying to get your foot in the door with a major retailer, a trade show is your best bet for making it happen. Bring that sell sheet with you.

DON'T

SPAM PEOPLE Direct email is a powerful tool, and you should start a list of customer email addresses on day one. But don't overuse it: Every time someone flags one of your emails as spam, your whole list loses value.

APE YOUR COMPETITION It helps to know where and how comparable products are being sold, but avoid using the same cookie-cutter tactics. Your goal is to stand out—not disappear or (worse) look like a knockoff.

SPREAD YOURSELF THIN It's usually cheaper to make new customers than it is to make new products. The success rate for new products is about one in seven, so only start developing a new one if your business is going strong and if you can afford to absorb a failure.

NEGLECT PLANNING It's easy to get trapped spending most of your time putting out fires or chasing the next goal—and very little time crunching numbers and thinking about the big picture. Resist with all your might. Winning startups spend more time strategizing than executing.

184 BE SEEN IN THE RIGHT PLACES

Marketers often use the word *channel* to describe how end users discover and buy a product—be it a point-of-purchase display at a big-box store or the catalog in the back pocket of an airplane seat. To find the right channel, think about where your customers hang out and how to get noticed there. For small manufacturing startups, the Internet has opened up huge opportunities for finding and selling to niche markets at very low cost—you see this on Etsy, eBay, and even Amazon. It's also possible to have success with your own independent ecommerce site. A good general strategy is to sell directly to end users through some online channel from the very start, then work toward growing your business so you can sell in bulk to established online, catalog, and brick-and-mortar retailers.

185 SELL YOUR PRODUCT AS SEEN ON TV

Whether it's Ginsu knives, the George Foreman Grill, or the unsubtle Shake Weight, everybody's got their favorite As Seen on TV (ASOTV) product—or at least their favorite to make fun of. *Direct-response television* (DRTV) is as old as TV itself, but its heyday began in the mid-'60s with Ron Popeil's Ronco brand, a line of kitchen gadgets that soon expanded to novelty devices for the whole household. In 1984, deregulation of the broadcast industry lifted federal restrictions on the length of television ads, and the 30-second Ronco-style spot ballooned into the 30-minute infomercial we still see on late-night cable today.

"It slices! It dices!"

DRTV is a specific type of *direct-response marketing*, which basically entails ads that require you to "act now" and contact the manufacturer directly to buy, with follow-on order fulfillment by mail. Even in the Internet age, this approach actually offers a number of advantages to both buyer and seller. "Cutting out the middleman" and "passing the savings on to you" are not just hype—by skipping over distributors and retailers, DRTV marketers really *can* afford to offer you a better product at a lower price than you're likely to find in a store. And though there are certainly rip-offs out there, a surprising number of ASOTV products end up having real staying power. Ginsu knives are still manufactured today, and for five straight years have been ranked by *Consumer Reports* as a "best buy" among 50 prepackaged knife sets.

"But wait, there's more!"

Advertising on TV means you know exactly when your ads will run, and handling all your own sales means you can directly connect every sale with a particular ad. That kind of detailed feedback makes it easy to tell quickly and accurately which ads, time slots, and products are working out. You can fine-tune your efforts using *A/B testing* (also called *split testing*), which entails running two different commercials to see which one generates more sales among your target audience.

186 PROMOTE ONLINE

The question, especially for small businesses, is not whether you should promote online but what tools and strategies will pay off.

SET UP A SITE Hosting a webpage where people can discover and buy your product is a no-brainer. Today, services like Squarespace and WordPress make it painless to maintain an online presence. If you can't fulfill orders, stock at Amazon, sign up for its Associates program, and put a widget on your site to direct consumers to the Amazon product page. (This will also kick cash from each sale back to you.)

SAY HELLO TO SEO Don't bury your invention five pages deep in search results. Include *keywords* (terms people will likely use when looking for your product) in your website's titles, metadata, tags, URLs, and image names. Adding a site map will help *spiders*—'bots that crawl the web in search of new content— find and boost you in search result rankings, as will frequently linking to your site from social media. Avoid Flash pages and text in your image files; spiders can't search 'em.

TRY ONLINE ADS To start, set up a Facebook page for your business and use its internal system to place ads, run promotions, and monitor interest. But don't stop at social: Look into *cost-per-million* (CPM) *ads*, for which you pay a flat fee for 1,000 *impressions* (the number of times an ad is shown to a user), and *pay-per-click* (PPC) *ads*, which you only pay for when someone—wait for it—clicks on the ad. For PPC, your cost is determined by the value of the keywords you're using. Hiring an ad network to help target your ads is usually money well spent.

187 SEE THE FLAWS IN THE JEWEL CASE

In 1981, the first compact disc rolled off the conveyor belt. (It was ABBA's *The Visitors*, for all you trivia-night types.) Audio industry giants Philips and Sony had been working for a decade to turn the optical work of American inventor James Russell into a commercial product. The task of designing the CDs' packaging landed on the desk of Dutch industrial designer Peter Doodson, who came up with the now-familiar hinged design that snaps together from three plastic pieces. The first samples out of the molds were so flawless that the name *jewel case* stuck. As Doodson later recalled, "At that time, they were not thinking of making a lot (. . .) It was expected to appeal to the audiophiles and sell in relatively small numbers."

Two decades later, though, billions of jewel cases were being manufactured every year. The design attracted widespread criticism, not least because it was prone to breaking and made of plastics that couldn't be recycled. Philips has

pointed out that the original weighed a sturdy 3½ ounces (100 g), and that breakage wasn't a problem until market pressure drove the weight down to 2½ ounces (68 g).

But there were other complaints. Fifty jewel cases take up 1 foot (30 cm) of shelf space, whereas the same space accommodates 200 CDs in cardboard sleeves. The cases are also slippery and difficult to stack or carry in groups, and come wrapped in a clear plastic that's a pain to open. Plus, they have an annoying antitheft sticker across the top edge that's tedious to remove and completely ineffective, since it's easy to pop out the case's bottom hinge, remove the CD, and close up the case again, all without damaging the sticker.

The point is that packaging matters. It's the first element of a product that consumers interact with, so it should be intuitive to use, easy to recycle or store, designed with minimal waste, and never a cause of frustration.

188 PITCH YOUR PRODUCT TO INFLUENCERS

In the Internet age, individual people—whether they're prominent bloggers, popular podcasters, social media celebs, or YouTube rock stars—can command an amazing amount of attention. Truth is, every hobby, interest, demographic, and lifestyle has its own online communities and influencers, and attracting their attention to your product can be one of the most cost-effective ways to generate publicity and sales.

STEP 1 Target the right folks. Hopefully you know enough about your customers by now to know what circles they move in online. If not, spend some quality time figuring out who the celebrities are in your space. Draw up a list of names and find whatever contact info you can. If you can't get an email address, send a message through Twitter, Facebook, YouTube, or whatever platform they broadcast on.

STEP 2 Make their jobs easy. Take high-quality, high-resolution pictures of your product against a neutral background and make them available online. Write some catchy copy describing who you are, what your product does, and how you're different from the competition. Make a bulleted list of your product's key features, and make sure you've got a URL where people can buy it with absolutely no frustrations.

STEP 3 Offer them freebies. Contact your influencers through the best means you have available. Tell them you'd like to send them a free sample of your product, or offer to provide products for a giveaway. (People love free stuff, and the influencers get a traffic spike out of it.) Ask them to reply with a shipping address if they're interested. If they don't get back to you, don't bug them. They can hurt you as much as they can help you, and if you're a nuisance they may well decide to.

STEP 4 Follow up politely. Ship your freebies with package tracking and, as soon as you see that the goods have been delivered, send follow-up messages including the high-resolution photos, your sales URL, and a polite invitation to contact you with any questions or comments. Including a personal phone number is a nice touch—these folks probably won't use it, but it shows that you're serious. Again, don't bug them, even if they just take the freebie and leave you hanging.

189 UNDERSTAND PACKAGE DESIGN BASICS

Packaging is its own product. There may be more or less hyperbole in that saying depending on your market and channel, but the point—that the box can be just as important to making a sale as what goes inside it—is demonstrably true, especially if you're selling in a retail environment.

In the case of bulk goods (such as hair gel and breakfast cereal), customers often choose one product over another based entirely on the package, sometimes even consciously and deliberately. In the gadget world, companies like Apple have elevated packaging design to a kind of cinematic art form, providing carefully choreographed unboxing experiences that guide the buyer seamlessly and intuitively through opening the product and using it for the first time. In some cases (like aerosol spray cans), the packaging is an inseparable part of the product itself.

There's obviously no one-size-fits-all approach. Package design begins by considering the contents and evolves to fill a series of increasingly complex demands. At the most rudimentary level, packaging must contain the product and any accessories in a single salable unit, and must protect the contents from damage before, during, and often after the sale. It must also identify its contents, if only with a UPC or other SKU label for stock-keeping purposes (and for mail-order-only products this may be enough).

If you're selling off a shelf, however, the box also needs to attract the eye, provide enough information to answer questions and close the sale, and hopefully serve the bigger-picture interests of building brand identity. To top it all off, today's savvy consumers are increasingly concerned with life-cycle factors in packaging, including environmental footprint, reusability for storage, and repackability for return or resale in the used market (see #140–142). And all these goals must be carefully balanced against costs—overdesigned packaging costs too much and eats away at margins, while underdesigned packaging saves too little and eats away at sales.

Finally, it's great if it's clever. Some see innovative packaging design as an indicator that there's a game-changing product inside. And updating the box around a familiar staple—like eggs—can make the same product seem fresher than the competition.

190 DISCOVER DIFFERENT PACKAGING TYPES

When it comes to bundling up your goods, you've got a lot of options, which typically fall into two categories. *Primary packaging* is the stuff that immediately covers the product at its point of final sale. It often features distinctive graphics to attract interest, inform consumers, and build brand identity, but it can also include filler to pad out and protect the package's insides. *Secondary packaging*, meanwhile, is the stuff that contains the primary packaging, either singly or in bulk, to protect it during transport. Here are some industry-standard methods that you might consider when designing your primary packaging.

BLISTER PACK

A blister pack is a sheet of formable material, often clear thermoplastic, with a depression or recess shaped to hold the product, sealed with a plastic or paperboard backing that often carries all the package graphics. It allows customers to get a good look at what they're buying.

CLAMSHELL

Any primary packaging that consists of two halves that hinge or fold along one edge to enclose the product is called a clamshell. These packages are often made of clear plastic with labels or inserts to carry graphics, and for durable goods they are often heat-sealed along the edges. They can be a pain to open. Make yours otherwise.

TUCK-END CLOSURE

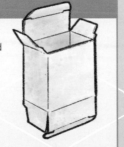

This box or carton end is templated and cut so it can later be assembled without adhesive or tape—just fold it up and you're done. It speeds assembly by requiring no tools, but it may not be as cheap, durable, or leak-proof as a sealed-end design.

SEALED-END CLOSURE

This box or carton has one end that requires adhesive or tape to close. It's more economical in terms of materials usage but it also requires operating a tool to seal, which may slow assembly. The sealed-end closure is stronger and more leak-proof than a tuck-end design. It's often favored for secondary packaging.

INSERT

Often made of folded paperboard or foam, an insert is an additional packaging structure that goes inside an outer package to restrain, segregate, and cradle the various parts of the product. It's sometimes formed to fit specific components. People also use this word to refer to instructions or other documents included in the package, particularly in the case of pharmaceuticals.

MOLDED PULP

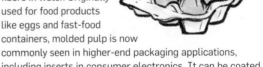

This recycled packaging material is made by casting a slurry of mashed-up paper fibers in water. Originally used for food products like eggs and fast-food containers, molded pulp is now commonly seen in higher-end packaging applications, including inserts in consumer electronics. It can be coated to resist water.

PAPERBOARD

Don't let the name fool you. While these sheet materials look like paper, they're thicker, heavier, and more rigid than the usual stuff you put in your printer. They may consist of several layers or plies, and may be coated on one or both sides for decorative or protective purposes. Varieties include boxboard, kraft board, chipboard, and containerboard. There's also corrugated fiberboard, a sandwich of flat sheets glued to a fluted corrugated sheet that gives additional strength, rigidity, and thickness.

191 ACE YOUR PRODUCT'S PACKAGING

If you're kitting, selling at shows or pop-up stores, or making mail-order-only products, you can get away with using off-the-shelf packing materials and graphics that you apply yourself. But if you're selling in a retail environment where the packaging will have to compete for sales, you should consider hiring a professional custom packaging designer.

To design your own packaging, you need a computer, vector-graphics software, a scale, and a ruler. Naturally, you also need to have a sample of the product for which you're creating packaging, along with any accessories that come with it.

STEP 1 Write a design brief. If you're hiring a pro, you'll give this document to her so she knows exactly what you want her to do. Even if you're doing it yourself, writing a brief is still a good way to organize your thoughts and data. It should include the final name of your product and company, correctly spelled and punctuated, with fonts and logos, if those have already been decided. It should also include the delivery deadline, any marketing data or user stories you've gathered, examples of competitor art, and any details about what should go on the outside and how you plan to manufacture the packaging.

STEP 2 Determine protective packing. Do you need a foam insert to keep your product from getting crushed or battered? Even if your product is really tough (like stamped steel hardware), your packaging ought to be rattle-proof. Consider environmental hazards like *electrostatic discharge* (ESD) and moisture too—you may need to enclose electronics in antistatic bags, or include a "getter" or desiccant pack in case your product ends up sitting in a humid tropical warehouse for a year.

STEP 3 Calculate dimensions. Once you know exactly what's going inside, figure out just how large the outer box should be. It needs to be just big enough to hold everything but not bigger than it has to be, which will avoid adding unnecessary weight, volume, and cost. If your display system or retail partner requires a standard package size, figure that out first and accommodate it.

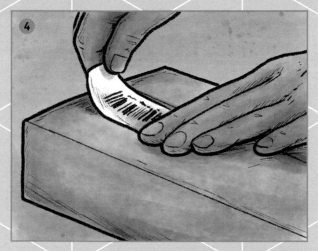

STEP 4 Figure out the fine print. Start with the essential UPC, SKU, or other stock-keeping bar-code label (see #171). Design and size the bar code to the standard for the scanning equipment you're using. Decide where it's going to go on the package, then start adding legalese: copyright notices, patent numbers, recycling instructions, warranties, and hazardous materials notices. Don't forget your website and a customer service number.

STEP 5 Unless you're selling something like coffee grounds, people expect to see what they're buying. Consider using transparent materials so customers can glimpse actual goods without opening the box, or put a professional-quality photo front and center. Add the product name and logo plus any features, claims, or hype to the front, giving these elements a clear visual hierarchy so they're read in the right order. On the back, provide more specific features and technical details, plus maybe a gettin'-to-know-ya story. When in doubt, look at your competitors.

STEP 6 Build and test a prototype package. Build at least one exact duplicate of the packaging, even if it means you have to cut out the box plan with a hobby knife and a ruler, then pack the contents inside exactly as they'll be arranged in the production packaging. Then heft it in your hand and see how it feels. Take it to the store and put it on the shelf next to the competitors, then stand back and look at it. Does it draw your eye? What questions do you have about it? Now pick it up and look closer. Does it answer your questions? Do you want to buy it? Ask friends, family, and strangers to repeat the same experiment, and get plenty of honest feedback before committing.

192 GET MARKETING TIPS FROM THE PROS

You've built it—now it's time to show it off to the world. Here's advice on getting your product out in the stores and into the hands of consumers.

"REAL-TIME MARKETING MEANS MOVING AT THE SPEED OF THE CUSTOMER, UNDERSTANDING THEIR PAIN POINTS OR INTERESTS, AND SERVING UP THE RIGHT INFORMATION AT THE RIGHT TIME." – *Karen Quintos, CMO of Dell*

"The best advice I can give to someone who wants to bring a product to market is go work for a company that does that. Spend a couple of years in someone else's employment learning how that whole world works. Then take your product to market. You'll have a much easier time." – *Maxwell Bogue, inventor of the 3Doodler, the 3D-printing pen*

On looking beyond marketing metrics: "We can't be obsessed or seduced by data. At the end of the day that emotional response is still a necessity." – Joseph Tripodi, CMO of Coca-Cola

"YOUR UNHAPPY CUSTOMERS ARE YOUR GREATEST SOURCE OF LEARNING." –*Bill Gates, founder of Microsoft*

"THE TRICK IS TO DISCERN A MARKET—BEFORE THERE IS ANY PROOF THAT ONE EXISTS." – *William Lear, inventor of the Lear jet*

"Invention is not enough. Tesla invented the electric power we use, but he struggled to get it out to people. You have to combine both things: invention and innovation focus, plus the company that can commercialize things and get them to people." – *Larry Page, cofounder of Google*

"VIRAL WORD-OF-MOUTH MARKETING FOR GOPRO IS MASSIVE. VIDEO IS REALLY THE CONDUIT." – *Nick Woodman, founder of GoPro*

"The most important word in the vocabulary of advertising is *test*. If you pretest your product with consumers, and pretest your advertising, you will do well in the marketplace." – *David Ogilvy, advertising executive that* Time *dubbed the "father of advertising"*

"Social enables word of mouth at an unprecedented scale. Its most powerful effect, through reviews and recommendations, is to put product quality and value for money as the key to success in commerce. Social brings a level of transparency that prevents marketers from advertising their way to success without underlying product quality." – Roelof F. Botha, venture capitalist

"PUT YOUR CONSUMERS IN FOCUS, AND LISTEN TO WHAT THEY'RE ACTUALLY SAYING, NOT WHAT THEY TELL YOU." – *Daniel Elk, cofounder of Spotify*

MEET MASSIMO BANZI
COCREATOR OF ARDUINO

Massimo Banzi comes from a family steeped in the manufacturing industry. His kin boasted an arsenal of practical skills that they applied at both work and home—a disposition that Banzi inherited. As a child in the city of Monza in northern Italy, the young Banzi often helped his dad fix up cars and spent his free time building his own electronics, including a radio. He also had a fondness for disassembly—to the point where others would sometimes bring Banzi things expressly to hack open. "When a kid starts to take things apart, I think that's an indication that his life might lead to tech," he says. Indeed, when he got his first computer, Banzi's interest in hardware expanded to software, and he taught himself to program.

Banzi studied electrical engineering but dropped out in 1992 and then began bouncing around Europe working on various projects, including as a software architect building big web applications and, briefly, as the CTO of an investment company. In 2002, he was invited to teach a two-week course at the Interaction Design Institute Ivrea in northern Italy, instructing master's students in how to use hardware and software. He liked the work so much that he decided to stay.

He faced a challenge, however: He needed to teach students—some of whom had never programmed or worked with electronics—how to tap into technological mediums, but he found that all the quality software and hardware on the market were quite complex for beginners to work with or else prohibitively expensive. So he decided to put together "a cocktail of different tools that would make the user experience very simple," as he puts it, and recruited several students and colleagues to help.

By 2004, he and his informal team had assembled the perfect kit, consisting of various

continued on next page

easy-to-use microcontrollers. Importantly, they kept the project entirely open-sourced. This allowed anyone who wanted to contribute to the platform's creation to do so, and it also created an economical option for those who didn't want—or couldn't afford—to purchase software or hardware, since they could simply make the products themselves. "We wanted this to be something that others could, in a way, own," Banzi says. "We're sharing the work with them, and they can benefit from it as much as we can."

The following year, they gave the project its now-famous name: Arduino, a hat-tip to Banzi's favorite bar in Ivrea, the King Arduino. While Arduino has since become one of the most popular platforms in the world among the maker community, in the beginning getting the word out was a challenge. Banzi and his colleagues spent years traveling around Europe, evangelizing about Arduino at universities, conferences, and gatherings. Some of their travel was sponsored by the places they spoke at, but much of it was paid for on their own dime. Slowly, though, interest began to build. A couple of schools adopted Arduino, and a demo at the Ars Electronica Festival in Austria provided an exceptional visibility boost. Mostly, though, it was just a matter of putting in the time needed to reach more and more people.

By 2009 or so, Arduino hit a critical mass, with sales at last allowing Banzi and his colleagues to start hiring people and "be more like a company." A few years later Banzi was invited to speak at TED—what he considers the company's final mark of success: "Once you get into TED, you've become mainstream."

Indeed, Arduino's website now receives around 25 million visitors per year, about two-thirds of whom are repeats. Each time the team releases a new version of the software, around 1.7 million people download it within the first 48 hours. The site also supports a lively online forum, where community members help each other troubleshoot and share pointers and projects. Users span cultures and continents, ages and genders.

"As the world becomes more and more complex and difficult, we want to make sure as many people as possible can participate in the process of innovation that goes on with technology," Banzi says. "It's a way to empower people and to reach a more diverse group."

Q+A

Q: What's your earliest memory of tinkering?

A: At eight years old I got a German set made by Braun called Lectron; it was electronic parts with magnetic tubes that could be snapped together. I built a radio, then from there I just kept building stuff.

Q: Why keep Arduino open-source?

A: We knew that it was impossible for the just the five founders—as we were at the beginning—to be able to do all the work that was needed. We wanted to make sure there would be collaborators who could join . . . We wanted this to be available in as many places as possible . . . If you cannot buy it you can make it yourself.

Q: What kind of customer support does Arduino have? If someone needs help, what are their options?

A: We have an online forum and people help each other there. If someone buys a board, we have support for them to get a replacement if theirs is defective, but if someone is building their own board they go to the forum and it's like peer-to-peer support.

Q: How do you think Arduino has affected the tech community at large?

A: There are a lot of problems with tech communities being dominated by white males, so we wanted to build tools that would increase the amount of people who participate in the innovation process. I'm happy that our community is doing a little more in terms of diversity, cultures, and genders. I'm glad to have helped improve things a little bit.

Since Arduino started producing microcontroller boards in 2005, they've continually iterated on their basic design and allowed others to develop their own boards using open-source hardware and software. Arduino has revolutionized the hobbyist landscape with its user-friendliness—just plug it into a computer via USB and load up your code, called a *sketch*, via Arduino's dedicated application (see #077). Arduino also offers a number of accessories and add-ons called *shields* that allow for further customization. Here's a peek at what's on the menu.

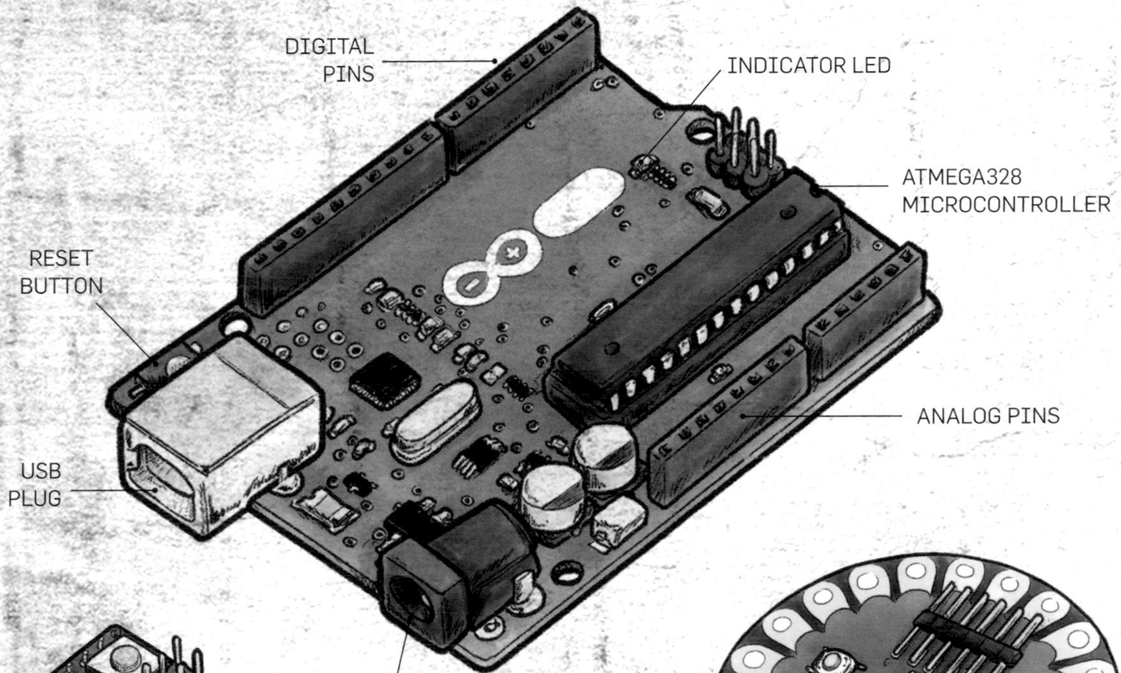

DIGITAL PINS

INDICATOR LED

ATMEGA328 MICROCONTROLLER

RESET BUTTON

ANALOG PINS

USB PLUG

EXTERNAL POWER SUPPLY PLUG

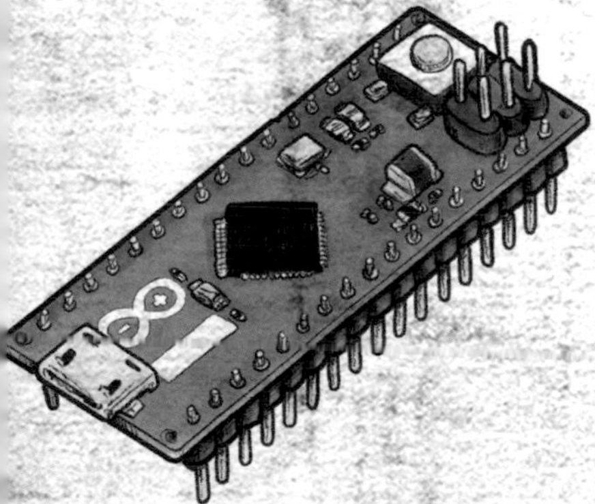

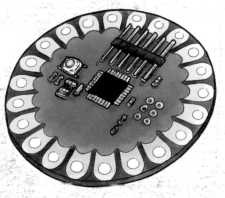

The aptly named Micro comes in at under 2 inches (48 mm) in length, making it easy to sneak into smaller everyday objects and run off a breadboard.

Developed by Leah Buechley for etextiles and wearables projects, the LilyPad is a lightweight board that can be sewn to fabric and attached to power supplies, lights, sensors, and more with conductive thread.

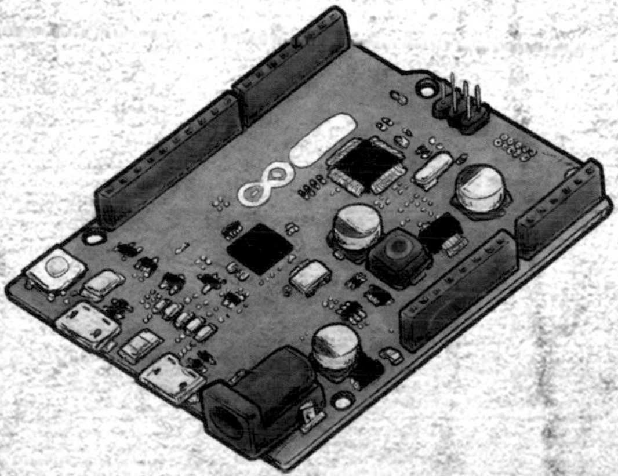

The Zero is Arduino's answer for more ambitious projects. Its increased power allows for Internet of Things, robotics, and wearables builds.

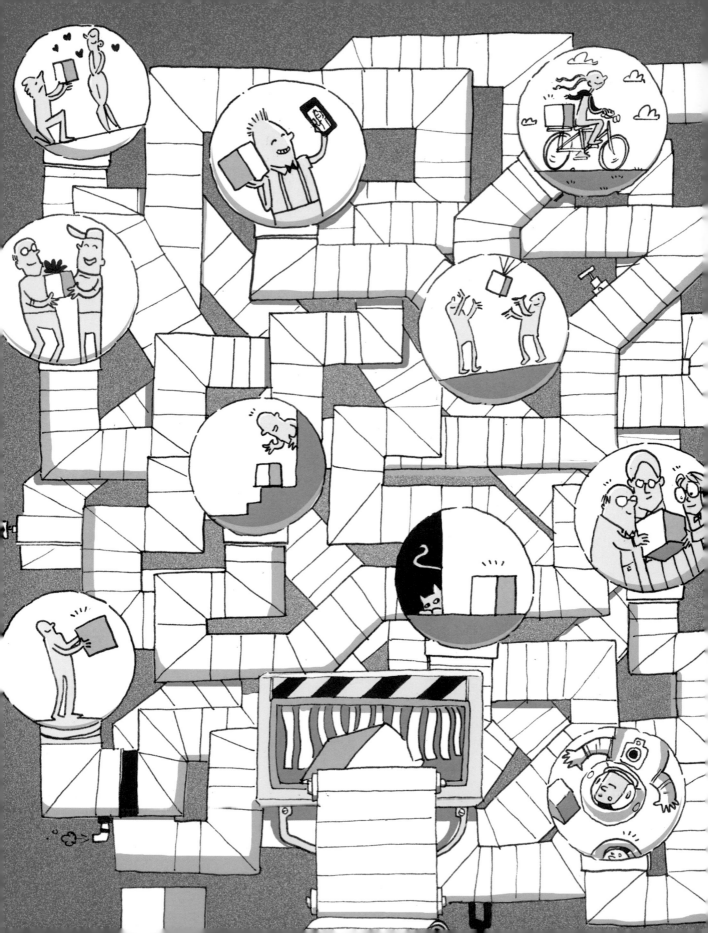

CUSTOMER SUPPORT

No matter how good a job you've done with design and documentation, you'll have at least a few customers who were expecting something else. And no matter how good your quality control, sometimes there'll be mistakes on your end. That's where the ongoing process of building a complete and rewarding customer experience comes in. Good customer service is essential to creating and maintaining repeat business.

194 DON'T BE THAT GUY

Customer service and *customer support* are phrases that tend to make a lot of us cringe, mostly because, as consumers, we've probably all had our share of frustrating experiences with bad customer support. Some of us may have worked in customer service at some point in our lives, and we largely tend to think of it as an undesirable and unrewarding job. And it seems like the bigger the company, the more confusing, tedious, and unpleasant the support experience is likely to be, for both customers and staff. As a startup,

at least, that's one thing you've got on your side: With a small team, there aren't layers of bureaucracy separating real customer interactions from the policy-makers. That doesn't mean that you can treat customer support as an afterthought—quite the opposite. Plan for and work toward achieving the best customer experience from the first day you turn on the lights. Or you risk falling into the all-too-common trap of treating these functions as an afterthought, thereby losing valuable business.

195 HELP THOSE WHO HELP THEMSELVES

Some customers will be inclined to solve their own problems. First, they'll play around with the product and try to figure it out for themselves, then they'll read the documentation, then they'll look around your website and maybe elsewhere online, and only as a last resort will they send an email or pick up the phone.

Make sure these people have all the best resources you can provide: Write good instructions (see #203–206) and provide a well-designed support interface on your website, including downloadable versions of in-package instructions, a continually updated Frequently Asked Questions (FAQ) document, and easy-to-find "real person" contact information, including an email form and a toll-free phone number.

Don't make people set up an account or a log-in to get assistance, and don't just dump people in a help forum with a search function and expect them to figure it out on their own. Help forums can be a useful asset, but they can also be a source of frustration. If you choose to implement one, make sure it's complemented by a clear, well-organized, and guided troubleshooting or FAQ page.

196 HELP THOSE WHO ASK FOR HELP

Unlike the more patient types who may be willing to spend time troubleshooting your product themselves, some customers won't be inclined to do this, and may be put off by any suggestion that they should have to. These people want to talk to you—right now.

As a cost-saving measure, many companies try to hide contact information for real staffers, or they build a funnel that forces customers to review self-help options before providing live support contact points. And when customer support becomes prohibitively expensive, this may be a regrettable necessity.

But the best approach prioritizes the needs of self-helpers and help-seekers equally, and recognizes that both are valuable. The self-helpers save you time, money, and energy by solving their own problems, while the help-seekers provide an opportunity to exceed expectations, create positive word of mouth, and provide precious feedback about how to make your product and documentation better.

197 HUMANIZE AND EMPOWER SUPPORT STAFF

Customers who contact you for support are, by definition, people who want to talk to a human being and who expect to be treated like one themselves. Don't make them interact with a machine—whether it's an email auto-responder, an automated voicemail system, or a person who's expected to act like a robot. Don't make them sit around on hold or answer a bunch of questions before being heard.

Ideally, a frustrated customer who wants to call can find your phone number painlessly, use it, and within 30 seconds hear a friendly voice answering with something like, "Hi, this is Barney with BlamCo customer support. How can I help you?" And then Barney listens and compassionately responds. He doesn't read from a script or use canned phrases.

If he can, he helps, and if he can't, he offers a refund, a credit, or a replacement product or part shipped free of charge. Or he offers to dig deeper and call back when he's found a solution. And it's only at that point—when it can be framed in terms of following up to fix the problem—that the customer's personal information is collected and preserved. To do those things, Barney needs to be empowered: He needs to be able to cancel the account, refund the money, fix the website problem, and basically do almost anything the CEO could do if one of her golf buddies called her private line to complain.

198 SET UP A CALL CENTER

Conventionally, a *call center* is a dedicated workspace for telephone operators to handle customer calls, either inbound (for support) or outbound (for sales). Thanks to hardware and software that make it easier to organize and analyze efforts, call centers have led the way in integrating computer and telephone technologies. For years, the call-center employee has worn a headset and sat at a desktop workstation that allows her to manage calls and access the company's customer-relations database through a single interface.

Today, however, many of the conventional ideas about call centers are fast becoming obsolete. Fewer customer interactions are over the phone, and more are by email, chat, and social media. Thus call centers are increasingly referred to as contact centers, and—as pressures to improve customer experience increasingly favor calling people back rather than leaving them on hold—the telephone operations are harder to classify as inbound or outbound.

Moreover, the widespread availability of connected personal computers has reduced the need for dedicated physical premises. Today, many operators work remotely using home equipment or even personal smartphones. This has effectively lowered the entry barrier for small companies who want to provide live customer support but might not be able to afford setting up a traditional call or contact center.

199 HEED THE TEN COMMANDMENTS OF CUSTOMER SUPPORT

You likely aren't permanently manning the phones of your company's customer support operation. But in the early days (or in a staffing pinch), you may find yourself wearing the headset—and hearing all sorts of questions and complaints. Take a deep breath.

THOU SHALT NOT TAKE IT PERSONALLY Customer support staff have to deal with human beings at their very worst, kind of like cops. Use a fake name if it helps you get into character, but don't let angry people get to you. It's not about you.

THOU SHALT LISTEN The story this person tells you is a valuable data point for your company. Pay attention and document it, and ask questions when you need to.

THOU SHALT CARE If you can't manage to find any authentic compassion for the customer you're dealing with, do your best to fake it. This is usually what angry people seek, and you'll be surprised how often even a sincere effort to empathize will calm them down.

THOU SHALT HEAR Writing down what they say and cooing at them isn't enough. You need to actively work to incorporate customer feedback through your support channels into quality control and product development. Use what they're telling you to make stuff better.

THOU SHALT TRUST Yes, there are people who will take advantage of you, but they're in the minority—certainly no more than 1 in 20. Unless their requests are so outrageous

that you really can't tell if they're serious or kidding, assume they're honest and make every effort to accommodate them.

THOU SHALT NOT ARGUE If necessary, explain—but don't quibble. Even if they bait you. The customer is not, in fact, always right. But they should always feel like they are.

THOU SHALT FOLLOW THROUGH If you tell someone you're going to do something, do it. Tie a string around your finger, put a sticky note on your monitor, or use a formal service-ticket software system. But don't break your word—no matter what.

THOU SHALT FOLLOW UP Nothing says you really care like a nice unsolicited email a week after a successful support interaction to make sure that the customer is satisfied. That's the kind of attention that creates lifelong loyalty.

THOU SHALT MAKE IT EASY Don't force your customers to work to get help. Contacting a real person at your company should be as easy for them as calling their mothers. Likewise, once they've contacted you, don't stall by making them set up an account, log in, or give you a bunch of personal details over the phone. Those are rage-inducing steps for anyone seeking assistance.

THOU SHALT HIRE PROFESSIONALS When you recruit customer support staff, don't treat them like interchangeable parts. Hire skilled customer experience professionals, put them on salary, and make them part of the team.

200
EXCEED EXPECTATIONS ON SOCIAL MEDIA

Having an online presence in the form of a dedicated product or company website is nonnegotiable these days (see #186). Likewise with social media platforms such as Facebook and Twitter.

It's best to batten down your URL and social media accounts early on. Otherwise, you run the risk of *social squatting*, in which someone claims your handle on prominent platforms and holds them ransom. (Just ask Morgan Freeman or Ray-Ban how annoying it is to get your handle back.)

Annoyances don't end there, though, as once you do have social media accounts, irate customers, crazy people, and your competitors can use them as a public form to air grievances, embarrassing you whether you deserve it or not. Some platforms (like Facebook) allow you to censor or delete negative comments, while others (like Twitter and Amazon) will not. In any case, reaching for the Delete button should be your very last option when dealing with a public complaint. From the start, cultivate an attitude that public attacks are actually opportunities to impress people with how great your customer support is.

Rule one is to respond both publicly and privately to every comment, if possible, and to do so as quickly as you can. Rule two is to surprise both the complainer and anyone else who might be reading with how compassionately and effectively you answer. From there, it's usually a good idea to take the conversation offline; it prevents escalation and promises a personal level of support. Don't you dare ignore these customers once they're in your inbox, however. You risk their renewed wrath.

201
LET SOFTWARE HELP MANAGE CUSTOMERS

Customer-relations management (CRM) is the effort to organize all your interactions with customers to improve their satisfaction and your business's performance. Today, CRM is virtually synonymous with the information technology—some hardware but mostly software—being employed in that effort. From a software perspective, CRM is fundamentally about building, maintaining, and using a database of customers and their interactions with the company, including purchases as well as personal contacts by phone, email, social media, or other channels.

For small businesses in particular, the most important function of CRM tracking is to make sure all the customers who should be getting attention actually are and to prevent wasteful duplicated efforts—for example, making sure one rep is not working to fix an issue already resolved by another. Higher-level functions include collecting and analyzing data about customers and their needs for use in setting policy, as well as automating feedback from customer service to engineering or other departments about bugs and feature requests.

CRM software options range from ad-hoc systems using cloud-based collaboration tools like Google Drive, to open-source code like SuiteCRM, to ready-made "freemium" programs like Insightly, to enterprise-level services like Oracle CRM and Microsoft Dynamics CRM. Though these programs are powerful, don't just throw down some money and expect your problems to be solved—think carefully about your current process, as well as how it's likely to change and grow, before choosing the tech.

202 DON'T CONFUSE SERVICE WITH SALES, LIKE COMCAST

Yes, providing consistent quality support can be a challenge. And yes, the more you grow, the bigger that challenge gets. Nonetheless, if you find yourself testifying before the U.S. Senate about your company's lousy customer service track record, you know something has gone very badly wrong. And that's exactly where Comcast Senior VP Tom Karinshak found himself in June 2016—standing tall before the Homeland Security and Governmental Affairs Permanent Subcommittee on Investigations, in a special hearing on "Customer Service and Billing Practices in the Cable and Satellite Television Industry."

To be fair, Karinshak wasn't the only one in the hot seat that morning. Alongside him were four more current and former execs from Time Warner, Charter Communications, DirecTV, and Dish Network. And while all five companies were called to account for chronic overcharging and pricing practices bordering on outright deception, Comcast was singled out for failing to resolve customer problems on the phone, subjecting callers to sales tactics, and making it hard to cancel or downgrade service.

In one infamous incident from 2014, tech writer Ryan Block published an audio recording from an 18-minute call to Comcast Customer Service in which an aggressive retention agent responded to patient and repeated cancellation requests with circular arguments and hamfisted efforts to change the subject. Then, in 2015, several Comcast customers who'd had unpleasant customer service calls later received mailings from the company addressed under vulgar epithets like "Asshole" and "Super Bitch."

Critics of Comcast's approach have repeatedly focused on the dangers of blurring the line between sales and service in customer relations. Rewarding service reps for upselling or retaining customers fosters interactions that can be very frustrating for both customers and staff alike and, especially in the age of social media, can generate lots of bad publicity.

HOW TO USE A FORK (NO, REALLY)

203 PLAN A KILLER INSTRUCTION MANUAL

We've all seen our share of lousy manuals: instructions with tiny unreadable fonts and confusing pictures, written or translated by someone with a comically bad command of the language. Such problems happen when companies treat this documentation as an afterthought or regrettable necessity rather than the golden opportunity it is.

Sure, instructions are first for people who already own the product and need to know more about how to use it. But don't forget that instructions are also of interest to potential customers who may base their decisions on nitty-gritty usage details they can only

find there. Smart companies treat instructions not just as a customer support tool but as a marketing one. (Anyone who doubts that statement need only watch the Virgin America instructional safety video.)

Whether you're designing a paper manual to go in a box or a set of webpages to support a software product, high-quality documentation can go a long way toward attracting and retaining customers, as well as building your brand's reputation through positive word of mouth. On the other hand, slipshod documentation can have the opposite effect. Take the time, sweat the details, and get it right.

204 SHOW DON'T TELL

In the instructional context, a picture is worth a lot more than a thousand words. Especially where physically complex assembly, repair, or use procedures need to be explained, no amount of text is a substitute for clear pictures—use them instead of words whenever possible.

KEEP IT MINIMAL While ink is pretty darn cheap these days, you should design as though it were gold—maximize the amount of information communicated by every mark.

ADD DETAIL TO CLARIFY Make your pictures as simple as possible but no simpler. More detail in an area will naturally attract the eye—use that effect to your advantage. Leave out details where they don't matter and add them where they do.

THINK IN LAYERS It should be easy for viewers to tune out the parts of a page they aren't interested in. The simplest trick is to use different line weights for drawings, callouts, and text, so that, for example, a callout line is never confused for a drawing line. Color can also be useful for this purpose but may add to printing costs.

USE MEANINGFUL LABELS If you're going to label parts in a diagram, use clear, concise names that make sense. Don't make the reader go back and forth between arbitrary letter or number labels and a reference table to figure out that part A, for example, is the battery cover. Instead just put the words "Battery Cover" on the picture itself.

DON'T BREAK THE FLOW When you use a sequence of pictures or diagrams, make sure that the individual images don't float in space or in time—it should be easy, just by looking, to see how each picture relates to the one that comes before or after it.

205 WRITE SIMPLY AND CLEARLY

The most timeless rules for clear prose come from George Orwell's 1946 essay, "Politics and the English Language." These are words to live by for any writing task, but they're especially relevant in product documentation, where the goal is to communicate as clearly as possible, using as few words as possible. Orwell's original rules are shown at left; to the right, I've rewritten each rule to violate itself.

GOOD	BAD
Never use a metaphor, simile, or other figure of speech that you are used to seeing in print.	Never use a metaphor, simile, or figure of speech that's been run into the ground.
Never use a long word where a short one will do.	Never employ a polysyllabic construction where a monosyllabic construction will suffice.
If it is possible to cut a word out, always cut it out.	If it is possible to cut a word out, always be sure to cut it out of there.
Never use the passive where you can use the active.	The passive is never to be used when the active is possible.
Never use a foreign phrase, a scientific word, or a jargon word if you can think of an everyday English equivalent.	Never use patois, neologisms, or argot if you can think of everyday English equivalents.
Break any of these rules sooner than say anything outright barbarous.	Any of these heuristics should be disregarded, if, in the course of putting them through their paces, one is impelled to commit unpardonable stylistic faux pas.

206 MAKE AN INSTRUCTION MANUAL

The specific goals for your instruction manual will obviously depend on the details of your product. Common jobs for instruction manuals include identifying the product and any bundled parts or accessories, showing assembly steps, giving operating instructions, describing maintenance or repair procedures, and providing additional contact points for those in need of more help. Manuals are often also the best place to communicate product safety warnings, warranty information, and other critical legalese.

As far as tools and materials to get started, you'll need a computer, printer, paper, pencil, and tape as well as three kinds of software: 3D CAD, vector graphics, and desktop publishing.

STEP 1 Identify your audience. Think in terms of demographics: age, education level, and so on. Remember also that not everyone who buys and uses your product is necessarily going to speak the same language as you. One approach to the language problem—exemplified by LEGO—is to use instructions that consist of mostly or even only pictures. Another is to provide different translations of the instructions depending on where you're shipping product. And a third, often the most popular, is to produce a single instruction manual in multiple languages.

STEP 2 Write an outline. List your specific goals for the manual, check it twice, then put it in order from highest to lowest priority. Use this list as the skeleton of an outline, with each goal becoming a top-level section. Now go back and start putting meat on the bones, adding specific details to each section of your outline until you're sure you've covered everything you need to say.

STEP 3 Figure out what form the instructions will take. Will they fit on one sheet of paper, or will they need a small booklet? What size page do you need? Will you print on one or both sides? If you're writing online or software-based instructions, how big is the screen likely to be? Is it more likely to have a portrait or landscape aspect ratio? Obviously, you have to balance the amount and quality of content against the cost of making the manual. Prepare a template in your page-layout software that matches the manual's final form, print out a set of pages, and tape them up. Now grab a pencil and start marking out where each item on your outline will go.

STEP 4 Make the pictures. These can be photos, so long as they're clear and of a professional quality. But if cost makes conventional black-on-white printing the most attractive option, go with line drawings instead. Fortunately, in the age of inexpensive 3D CAD software, you can produce high-quality perspective line drawings without any special artistic ability: Just set up the view you want in your CAD program and export vector art, if that's an option,

or a raster image if it's not. If you can only get raster images out (even if they're only screencaps), it's pretty easy to trace over them by hand in a vector-art program.

STEP 5 Fill in the words. Once all the pictures are in place, go back and add any words that you absolutely can't avoid, using as few of them as possible. If your product is intended for children, don't write over their heads, and remember that adults also have widely varying literacy skills. Every word you type may need to be translated into one or more different languages, depending on where your product might someday be distributed. If writing isn't your forte, consider hiring a professional, and absolutely do not depend on machine translation.

STEP 6 Get some feedback. Before committing to a print run, mock up a few samples and pass them around to friends, family, coworkers, and maybe a few strangers bribed with Starbucks gift cards. If you've included safety warnings, terms of service, or other legal stuff, have your lawyer sign off on it too. A professional copy editor—or maybe a college student enrolled in an English or journalism program—might also be good to tap to ensure the manual is clean. And again, pay extra attention if you're working up foreign-language translations. Even if you paid a professional translator or translation service, it's a good idea to find at least one more native speaker and get her take on the translated copy.

207 BALANCE ONLINE AND IN-THE-BOX HELP

It's tempting to skip printing a paper manual altogether and instead direct users online for instructions. This is usually a mistake. When in doubt, fall back on these three rules of thumb.

MAKE IT SELF-CONTAINED There are admittedly many advantages to putting a manual online: You save dollars and trees, and it's easier to update. The problem is when a customer gets frustrated when he needs help but doesn't have Internet access. That experience can bounce back on you as bad word of mouth.

MIRROR EVERYTHING ONLINE Whatever you put in the package should be freely available for download online, ideally as a PDF. This attracts

more meticulous customers researching specific details about your product before they buy. Also, many customers may not want to keep a paper manual but prefer to save a digital copy.

SKIP VIDEO Have you ever looked for an answer online only to discover that all the help content is buried in videos? Don't do this to your customers. Put all instructions in easy-to-find, easy-to-read text files.

MEET ERIC STACKPOLE
COINVENTOR OF OPENROV AND TRIDENT

While many inventors hope to strike gold, very few take the idiom quite as literally as Eric Stackpole and David Lang. The pair—one a former NASA engineer and propulsion expert, the other a self-taught sailing aficionado with marketing moxie—developed a submersible robot to search an underwater cave for hidden treasure. "The story goes that there was a gold robbery in the 1800s, and the thieves had to ditch the goods in an underwater cave because the weight was slowing them down," Stackpole shares. "Long story short, the sheriff's posse caught them, demanded to know where the gold was, and then hanged them on the spot. When they got to the cave, there was no technology to explore it, so they gave up."

Stackpole had actually been working on some version of his underwater robot for quite a while. He started tinkering with it as a teenager, and later worked on it in the evenings as stress release from grad school and his job building spacecraft. It wasn't until a friend tipped him off about the sunken loot, however, that he felt the urge to make his deep-sea dreams come true. "I suddenly had a mission, and once you have a mission, you have a purpose. Things accelerate a lot."

It was about this time that he met Lang, who saw the potential in Stackpole's completely open-source exploration tool. It was Lang's idea to set up a website and build a community around the invention. "At first it was just me and him talking back and forth to each other on the forums," Stackpole jokes. "But eventually we went to the cave and tried to fly our ROV there."

Spoiler art: It didn't work very well. But the *New York Times* picked up the story, and suddenly Stackpole and Lang had 70 people logged onto their forums. It was fuel for the fire. "In my experience, technological or capability barriers don't kill projects—motivational or attitude barriers do. But if you have people rooting you on, you're much more likely to succeed."

continued on next page

In addition to acting as a built-in cheerleading squad, the community opened up their wallets and spread the word when Stackpole and Lang launched a Kickstarter for their DIY kit in 2012. The kit was still in development, but they were honest about where they were in the process and invited the crowd to contribute ideas about usage and the tool itself. One community member posted about a more efficient way for data to travel through the tether (it involved using a specific type of home plug adapter) and ended up joining the engineering team. Scores more shared hacks to the robot, such as water samplers, hooks, and GoPro cameras to aid in *photogrammetry* (an imaging technique that helps in surveying topography). "It's a buffet," Stackpole says. "People can pick and choose the parts that they want, and then innovate and add on others."

Having a fanbase also helped Stackpole and Lang scale up when it was time. According to Stackpole, "We were the top-selling ROV manufacturer in the world, but we realized there must be an even larger number of people who would get involved if they didn't have to spend a few weekends building a kit." The pair had also learned a lesson from watching the rapidly changing drone market: The startup 3D Robotics had at first dominated with their kit, only to be overtaken by DJI, a Chinese company that began selling completed products. "They knocked the socks off everybody—they created a sensation because they had something that was ready to go," he says.

To capture this marketshare, the pair made Trident, a completely assembled underwater robot with improved navigational control and video quality. They got some input about what features to add: "I was on an Arctic expedition for three months, and I talked to scientists who wanted to fly their submersibles in long straight lines to survey an area, rather than doing pinpoint observations," Stackpole recollects. "We also heard that the camera really mattered, so we spent a huge amount of time and money making it send out beautiful videos." Then there was price: Most ROVs cost US$10,000, but Stackpole and Lang are committed to keeping their gadget as affordable as a laptop computer.

After all, their entire mission is about "getting more eyes in the water"—about democratizing exploration. "You don't have to get a research grant to pursue your curiosity," Stackpole says. "You can answer your own questions with the tools that are available today."

Q+A

Q: Are there any other inventors or makers who particularly inspire you?

A: Burt Rutan, the owner of Scaled Composites and the guy who did Space Ship One. One day I got to meet him and asked what I could do to be a part of this new era of exploration. He said, "Take your concepts, and build them, and fly them." I took that to mean, "If you have an idea, just start doing it." I live by that.

Q: What's it like to see your invention in the wild? Are you ever surprised by its uses?

A: Even in the early days, when we just had a few hundred OpenROVs out there, we were seeing footage from people doing something incredible with them. We had footage of someone who flew theirs in the South Pacific with pods of melon-headed whales. And we had footage of people flying them in Sweden, looking in these rivers that seemed pristine but once you submerged you saw that they were completely filled with trash—bicycles and shopping carts. These exploration stories were big motivational uppers.

Q: What's your dream invention—if time, money, or the laws of physics didn't apply?

A: If we're really going crazy, I'd love a "thrust box"—a box where you could put energy and get a lot of thrust out. It would basically be a propulsion system that could run off energy efficiently. You could have flying vehicles, space travel, all these amazing things.

Q: Do you have any words of wisdom for aspiring inventors?

A: Be genuinely interested in and passionate about the thing you choose to work on. Do that thing you think about in the shower.

Trident, the new and improved version of OpenROV, comes with a lot of great features that will help deep-sea explorers better navigate and see underwater, as well as gather vital telemetry about the world below. Recently, Stackpole and Lang demonstrated its prowess on a dive of the wreck *SS Tahoe*, a steamboat that sunk In 1940 to 500 feet (150 m) below the surface of California's Lake Tahoe.

Telecommunications, like live video and data, make their way back topside via a buoyant tether.

For high speed and easy maneuverability, Trident's special propulsion system includes three thrusters: The two horizontal ones move the ROV forward and backward, and from left to right, while the vertical one allows for up-and-down motion.

The streamlined shape of this underwater drone helps it move in a straight line for long distances. Its outer frame is protected by a rubber coating.

LED lights help brighten up deep, dark underwater areas. They also allow for fun night dives.

A high-definition camera relays live video to operators at the surface.

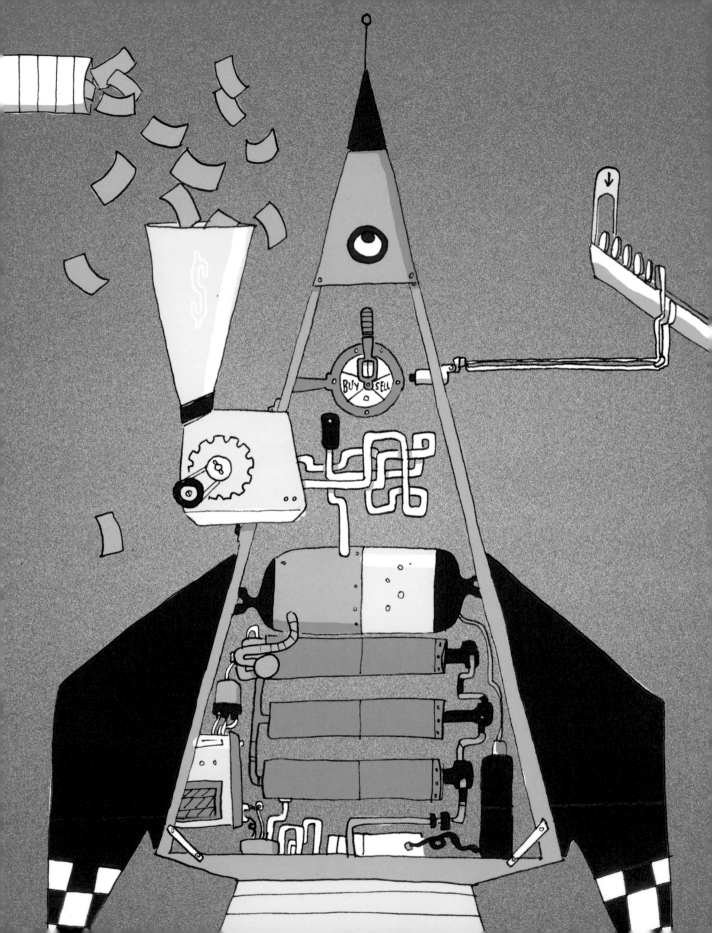

SELL OUT OR SELL ON

Building a business is a great personal achievement, and selling a successful one can generate great personal wealth. But whether you're looking to buy a yacht and sail to Ibiza or just cash out and move on to something shiny and new, the sale itself is obviously a critical process. So you should plan and manage this stage very carefully to make sure you get every dollar you deserve.

209 PICK YOUR MOMENT

Entrepreneurship is about more than just making money. Maybe you like the independence of working for yourself, maybe you're passionate about your product, or maybe you want to build a family business to hand down to the people you love. Still, the decision to sell or not is incredibly complex, and it depends on many factors, some of which are entirely outside your control—an economic downturn, the entry of a strong competitor, a critical technology shift, or a regulatory or geopolitical change.

You certainly don't have to sell if you don't want to, so long as you remember that choosing not to sell can have costs too—it might mean missing a key window of opportunity and could be worse than selling too soon. To avoid missing your moment, pay attention: Develop your own personal intelligence network that keeps you apprised of what's going on in the world—on Wall Street, in Washington, in your industry, in your hometown, and in your organization.

And while it sounds like common sense, it's crucial to pay attention to what matters to you. Are you still passionate about the project? Does running it help you further your goals, or do you feel yourself drawn away from your core interests? Do you have a new financial risk in your life that might justify a sale? Hear yourself out, as health and happiness matter more than money.

210 LOOK FOR A HIRED GUN

The art of buying and selling companies is known as *mergers and acquisitions*. Patrick Bateman, the businessman and serial killer of *American Psycho*, commonly confuses this phrase, which describes his day job, with *murders and executions*, which is how he spends his nights. An extreme image, yes, but a memorable caution for anyone who ventures into this world: Those who thrive here are sharp, to be sure, and you have to be sharp yourself to avoid getting taken.

If you're the type who can sit at a card table and know within a few hands who the sucker is, you might be able to hold your own, but experience in selling companies still counts for a lot—this is one of the most complicated transactions that exists, after all, vastly more complex than selling a home. If you have no experience and don't feel very confident, consider hiring out the acquisitions process to an expert. A good M&A adviser can mean the difference between selling and not selling, as well as millions of extra dollars in your pocket if you do sell. Balanced against the potential risks and rewards, the cost of hiring an adviser is small and likely to be well worthwhile.

211 PREP BEFORE SELLING

Unless you're a regular big shot on the poker tourney circuit, the negotiations to sell your business will be one of the biggest gambling matches of your life. Here's how to ensure you'll be satisfied when it comes time to cash in your chips.

KNOW YOUR BEST ALTERNATIVE Before you even start talking to potential buyers, decide what you're going to do if you can't sell your company at all. More than just having a vague idea of an alternate plan, know how much money you can expect to put in your pocket in that case. Unless you know that number, you can't make a rational choice about when to say no.

KNOW WHO'S THE BIGGER FISH In a *merger*, both companies have approximately equal power, meaning the same value to gain or lose by the transaction. Called a *win-win negotiation*, it's a lower-pressure situation. In an *acquisition*, the acquired company is at a power disadvantage, and things are more antagonistic: Your goal is to get the highest price for your company, while the other company's goal is to get it for the lowest price. This scenario, called a *win-lose* or *zero-sum negotiation*, is more likely.

KNOW YOURSELF If you're being acquired, your goal is to persuade the other side that your company has the highest possible value. That means not just naming an aggressive number but having arguments, data, and evidence to back that number up. Be ready to explain why your number should be as high as it is and why lower offers are unfair. Also be quite sure about your *reserve price*, which is the worst deal you're willing to accept.

KNOW YOUR FRENEMY Learn everything you can about the acquiring company and the particular people you'll be dealing with. Get creepy about it: Look them up on Facebook or other social media platforms, spend a few dollars to get background and record checks, and basically absorb every bit of information you can. It's crucial to understand where they're coming from, why they're interested in your company, and what their best alternative is if the sale doesn't go through.

212 PERFORM A SWOT ANALYSIS

SWOT stands for *strengths, weaknesses, opportunities,* and *threats,* and it's the standard model for summarizing a proposed business venture. When you start reaching out to potential buyers, it's wise to include a candid SWOT analysis of your company in the introduction of an information packet (see #222).

STRENGTHS These are the things your company does well. Got an especially strong brand? A really high-margin product? A network of valuable relationships within your industry? This is the place to brag about it.

WEAKNESSES These are the things your company could be doing better. Do you only have one or two products? Problems with cash flow? A small independent team with problems adapting to a big corporate bureaucracy? Own up to it here.

OPPORTUNITIES These are strategies your company could use to grow. Is your market getting bigger? Do you have a newly issued patent? Are laws governing your industry about to change in your favor? Paint your rosy picture of the future here.

THREATS These are possibilities that would destabilize or devalue your company. Facing lots of strong competition? A hostile regulatory environment? An economic tailspin that threatens your margins? Don't gloss over these aspects.

213 SET THE BEST VALUE

The price is the number that goes on the contract, and it depends on the outcome of your negotiations. The value, however, is different—it's what your company is intrinsically worth, and it's trickier to pin down. So your goal will be to determine a value range rather than a specific number. Even this is a very complicated problem that likely justifies hiring an expert.

The valuation process involves weighing the expected profits from your company against the risk that it won't meet those expectations. There are several methods for doing so, and you should choose one that establishes a range that feels right to you as the owner. (Similarly, you should expect the buyer to choose a valuation method that feels wrong, setting the range as low as possible.) One common method is known as *discounted cash flow*, which projects a company's expected cash flow over the next five years and applies a discount rate to account for inflation. Another tactic is based on revenue: Multiply your annual income by a risk factor to arrive at a snapshot of how your business is expected to perform compared to others. The risk factor is determined by considering stuff like the overall health of your industry, the predictability of sales within it, and red flags like unprotected IP and equipment in need of maintenance or upgrading. You can use a similar method starting from profits, rather than revenue; this works well when your margins are high. Agreeing on a price somewhere between the buyer's and seller's valuations is the point of negotiation. Avoid setting an arbitrarily high valuation as a negotiating tool unless you're prepared to defend it in convincing terms.

214 FIND THE RIGHT BUYERS

The trick about value is that it doesn't really exist apart from a particular owner. Just as one man's trash can be another's treasure, your company can have a very high value to one buyer and a very low value to another. That's why finding the right buyer is probably the most important part of the sales process.

If you're approached by a buyer, it's a good sign that your company is valuable to him. Keep him talking while scrambling to understand why he's interested. Identify his competitors, and understand that these companies may also want to acquire you. Having multiple interested buyers is almost always in your best interest—it gives you negotiating options and opens up the possibility of selling at auction.

That's no less true if you have to go out looking for buyers, though the approach has to be handled carefully. When combing for potential buyers, think first about the companies you do business with every day—your suppliers and your clients. Either group might be interested in growing by vertical integration. Foreign companies in a similar business may want to acquire you in order to expand into a new territory. Your direct competitors may also be interested, though these need to be approached from a position of strength.

215 NEGOTIATE LIKE A PRO

For most people, negotiation is not fun. It may bring up memories of haggling over a used car, or trying to get up the courage to ask for a raise. But, as usual, the only thing to fear is fear itself. Here are some tactics for surviving a negotiation.

STEP 1 Get them to the table. If the other side is digging in against negotiating, try putting a price on inaction. Perhaps start courting another buyer.

STEP 2 Serve food. Food-sharing is a deep, primal bonding ritual for human beings. Parents share food with children. Couples share food on dates. Family and friends share food on holidays. Take advantage of these norms to establish a comfortable atmosphere.

STEP 3 Some negotiators may ask you to give a range of prices you'd accept—i.e., revealing your reserve. Don't fall for it. Crack a joke, change the subject, or stall for time to think it over.

STEP 4 If you have a good idea of the most the other side is willing to pay, it can help to name a price first, which should be above that maximum but not so far above as to be outside the possibility of negotiation. This is called *anchoring*.

STEP 5 Counteranchor. If you don't have any idea of their maximum price, wait to let the other side anchor. If their offer is dramatically higher than you expected, congratulations, but keep a poker face—you need to make them think that's about what you were expecting. If it's dramatically lower, don't counter with your own number immediately. Instead, start building your case based on your valuation research, then present your price at the end, as a logical consequence of your data.

STEP 6 Concede reluctantly. Since negotiating is uncomfortable for most people, there's a natural pressure to get it over with quickly. This can leave money on the table. Large concessions send a message that you're willing to go further, whereas small, slow concessions signal that you're not going to budge. Stall for time before giving ground.

STEP 7 Expect exploding offers. Smart buyers usually put a time limit on their offers. If you receive an open-ended offer for your business, recognize it as an advantage: You can look for a better deal elsewhere and still have the original offer to fall back on.

STEP 8 Overcome the siege mentality by offering the buyer a way out: You can buy lock, stock, and barrel at full price, or you can buy everything except such-and-so at a lower price. If your assets include a patent, maybe lower the price in exchange for a royalty or licensing deal.

STEP 9 Seal the deal. Negotiations should wind down naturally, with everyone understanding the outcome. If you've come to terms, put them in writing, then have everyone sign.

STEP 10 Resist backsliding. A signed statement of terms, though not legally binding, is a powerful disincentive to ill behavior. If someone tries to renegotiate a done deal after the fact, make it clear that changing any terms will require renegotiating the whole deal from the start.

216 STUDY UP ON MASSIVE IPOS

The bell seemed to ring especially loud the morning that these companies debuted on the floor of the New York Stock Exchange. Here's how they positioned themselves to succeed in the market.

US$358 MILLION

FITBIT This personal-fitness tracker achieved success with a hardware-software double whammy: a stylish, discreet device and a data-rich, socially motivating app.

US$2.9 BILLION

SQUARE Silicon Valley was beyond excited about the IPO of this portable credit-card reader, which revolutionized how independent vendors do business.

US$3 BILLION

GOPRO The stock price of everyone's favorite action cam jumped more than 100 percent during its first week of trading.

US$25 BILLION

ALIBABA The king of all IPOs, this Chinese ecommerce company built a global business-to-business network to connect and serve small enterprises everywhere.

US$584 MILLION

SOLARCITY While investors had been burned by solar-panel manufacturing before, Solarcity sidestepped the issue by offering installations and financing instead.

US$16 BILLION

FACEBOOK Many were disappointed on the first day of trading for the social media giant, but those who kept the stock saw it soar by more than 105 percent.

217 DECIDE IF YOU SHOULD STAY OR GO

It takes guts and grit to start a company—to stick with an idea through all the many stages of development, and then remain at the helm once the more ho-hum operational duties replace the hands-on work of inventing. If you get the itch to sell, ask yourself the following questions.

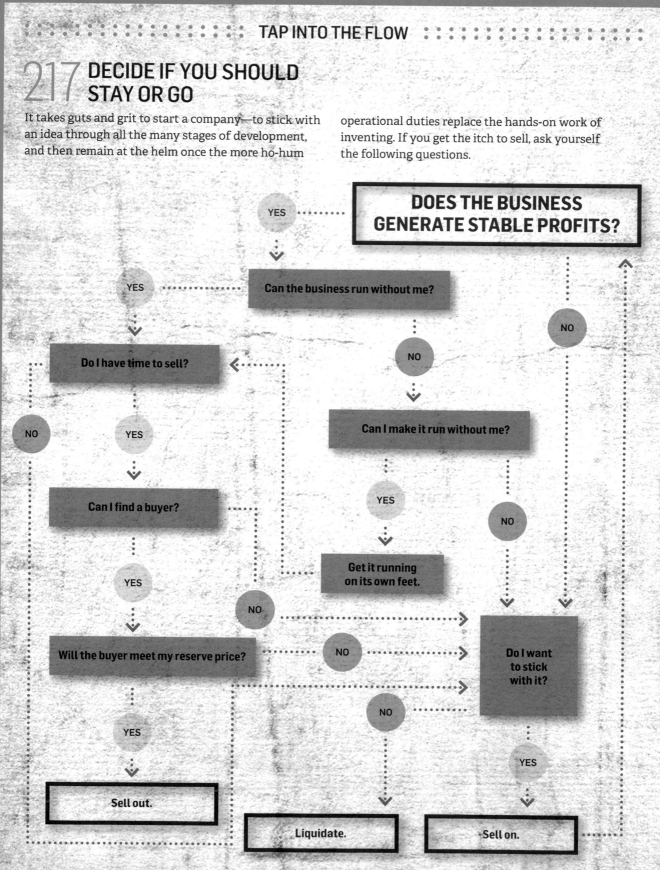

DOES THE BUSINESS GENERATE STABLE PROFITS?

YES → Can the business run without me?

YES → Do I have time to sell?

NO

YES → Can I find a buyer?

YES → Will the buyer meet my reserve price?

YES → **Sell out.**

Can the business run without me?

NO → Can I make it run without me?

YES → Get it running on its own feet.

NO

NO → Do I want to stick with it?

NO → **Liquidate.**

YES → **Sell on.**

218 CASH OUT WITH AN IPO

Making an *initial public offering* (IPO, aka going public) means changing the ownership of your company from privately held to publicly traded. Shares of a privately held joint-stock company are frequently offered for sale first to institutional investors like banks, hedge funds, insurance companies, and pensions in what's known as a *primary offering*. These institutions then relist the stock on public exchanges such as the New York Stock Exchange, NASDAQ, and the London Stock Exchange Group, where it can be bought and sold by anyone. That phase is called a *secondary offering*.

An IPO often raises operating capital for the company, but it also commonly serves as a way for the owners to cash out and convert their interests in the company into personal wealth. Other advantages include raising the company's public profile and attracting better talent through equity-based incentive options. Disadvantages include loss of control by the original owners, greater exposure to legal and financial risks, and time and money costs associated with receiving an offering.

Going public tends to be a thing of prestige—it usually indicates that your company has made it and is poised for growth. As such, the barrier to entry tends to be high: For instance, the New York Stock Exchange and NASDAQ both require at least US$10 million in pretax earnings over the three years prior to filing, which is a lot of clams. It also requires a strong management team—not just smart folks who show up every day with good ideas, but leaders who can speak to the company's vision and interface with investors. Last but not least, it's crucial to have the right investment bankers at your table. You want expert analysts who understand your business and industry.

219 GIVE LICENSING A GO

While selling your company for a cool million or two is the dream, you have other options here—like licensing, which allows you to collect royalties while benefiting from the licensee's resources and channels. For many, in fact, this is the easiest and least time-consuming route. The biggest downsides are that you'll no longer control your invention and you'll only receive a fraction of the profits. On the flipside, you won't be responsible for manufacturing, distributing, or marketing your product.

STEP 1 Research companies that may be a good fit. Would they be interested in your product? What's their reputation and track record? Compile a healthy list and rank it in order of preference.

STEP 2 Protect yourself. You need a liaison who speaks legalese and is invested in helping to secure the most favorable licensing agreement for you, aka a lawyer (see #095). You also need patents (see #106–111) and proper trademarks (see #096) to ensure your idea stays yours. You should also work with your lawyer to draft an NDA (see #223) for potential licensees to sign before you divulge the details of your invention.

STEP 3 Predefine what a good deal looks like. Your lawyer can come in handy here to help you determine what a reasonable royalty is for your product and for what length of time you want to license it. Remember, no deal is better than a bad deal. Knowing what you want out of the agreement is essential.

STEP 4 Create a product sell sheet that you can use to introduce your invention to potential licensees. Be sure to include the need your product meets, as well as its main features, target market, and legal status. Your sell sheet and prototype are your main tools to persuade your licensee to sign on.

STEP 5 Reach out with an introductory email or letter, along with your sell sheet. Work down the company list you made in step 1.

STEP 6 Meet with interested parties, present your product, and, if they go for it, discuss your ideal terms. Remember, you've put a lot of blood, sweat, and tears into your invention. Since you've come to the meeting with clear parameters of what would make the deal worth it to you, be firm and confident.

220 BE SAVVY ABOUT SELLING

Few transactions are as complex as selling a business. If you've ever sold a home, consider how much more involved it was than, say, selling a glass of lemonade. It's not much of an exaggeration to say that selling a home is like selling lemonade when you compare it to selling a company. The process can take as long as two years and rarely happens in fewer than six months; nine months tends to be about average.

Unlike selling a home, confidentiality is usually extremely important in selling a company. Rumors that a company is about to change hands can dramatically affect its value—for instance, they can trigger staff turnover, scare off clients, or encourage tactical behavior among competitors. Also, a company

that has repeatedly failed to sell or has been on the market a long time will be perceived as undesirable. (You know that one house that's always for sale? It's best not to be the business version of that.)

In general, you must balance two conflicting forces: the pressure to keep things on the down low and the desire to realize maximum value by having as many competing buyers as possible. If rumors of an impending sale could cost you a lot (if, say, you have a competitor closing in, or high-value labor that you can't easily replace), approach fewer potential buyers—perhaps even just one. But if you have fewer reasons to keep things under your hat, you could advertise the sale or even offer your company at auction.

221 | STUDY UP ON MERGERS AND ACQUISITIONS

When ownership of a company is transferred or combined, mergers or acquisitions result. Wrapping your head around the lingo will help you project a more professional vibe during negotiations.

MERGER

A *merger* or *consolidation* happens when two businesses of approximately equal value unite to form a single company. This unification helps both companies take advantage of economies of scale, expand their reach, or gain market share.

VERTICAL MERGER

A *vertical merger* is when one company unites with another along the same supply chain—perhaps one of its former suppliers or clients. You might, for instance, eventually decide to merge with a contract manufacturer who has been building your products.

HORIZONTAL MERGER

A *horizontal merger* is when one company unites with another along a parallel supply chain. For instance, you might eventually decide to merge with a company selling a similar product.

CONGLOMERATE MERGER

A *conglomerate merger* is when one company unites with another that's in a totally unrelated business. This strategy is sometimes called *diversification* and is intended to protect the conglomerate against instability in one particular market.

STRATEGIC MERGER

A *strategic merger* is when two companies have a long-term plan in mind when choosing to unite. A foreign company that wants to expand operations into a new country will often seek a strategic merger with a native company.

TAKEOVER

As opposed to a merger, a *takeover* or *acquisition* involves one business (usually of greater value) purchasing another. Sometimes it's smarter for the smaller company's brand and corporate identity to be preserved rather than the larger company's. This is called a *reverse takeover*.

PRIVATE ACQUISITION

A *private acquisition* is a takeover in which the acquired company doesn't have shares traded on a public exchange like the New York Stock Exchange or NASDAQ. As an inventor and small businessperson, it's most likely that your company will be sold as a private acquisition.

HOSTILE TAKEOVER

A *hostile takeover* occurs when one company purchases another, possibly against the wishes of management, by buying most of its shares. Unless your company is publicly traded, you probably won't have to worry about this.

MAJORITY STAKE

A person or corporation who owns more than 50 percent of a company's equity is said to own a *majority stake*. A majority stake grants a controlling interest in the company in question, meaning that a majority shareholder is in a position to wield significant influence over the company's actions and policy.

EQUITY PURCHASE

You may remember that a business's equity is equal to its assets minus its liabilities (see #985). Equity represents a business's monetary value to its owners and it may be divided among various owners by allotment of shares. An *equity purchase* involves buying up some of the ownership of a company, often in the form of shares.

222 SELL YOUR COMPANY

The process of selling a business typically involves offering more and more disclosure in exchange for more and more commitment from a potential buyer. Interested parties will naturally try to discover everything they can before signaling intent, and you've got to be diplomatic in managing these requests while being careful not to give away too much too soon.

STEP 1 Make a short list of need-to-know people who can freely discuss the impending sale—perhaps your partners, advisers, and key staff. Ask each person to sign an NDA (see #223 at right), then explain that you need their help in preparing for a sale.

STEP 2 Write two documents: a *blind teaser* and a *confidential information memorandum* (CIM). The first is a one- to three-page summary that presents key investment data without revealing your company's identity, letting you advertise clandestinely. The second is a document much like a business plan to show to potential buyers who've signed NDAs. It should professionally, accurately, and persuasively present the company as an investment opportunity without naming a price.

STEP 3 Contact as many likely buyers as you can. Prepare a list, including names, titles, phone numbers, physical addresses, email addresses, and biographical info of decision makers. Send anonymous blind teasers to these people from an email address that they can't connect to you or your business.

STEP 4 Meet and greet. The first contact should be by telephone from a third party shortly after blind teasers have arrived. Those who express interest sign an NDA and then receive a professionally bound copy of the CIM. Only then do you disclose your company's identity and meet face-to-face. Prepare your team for these meetings so that you can send a coherent message that's consistent with your strategy. Don't fall for ploys to get people talking separately.

STEP 5 Negotiate offers. At some point, you'll need to draw a line in the sand about how much information you're going to share with a potential buyer without getting an *indicative offer* on the table. This offer should include a price, financing details, a proposed timetable, and any conditions that the offer is subject to—for instance, approval of bank loans or the outcome of due diligence research. An indicative offer is sometimes called a *conditional offer* or *letter of intent*.

STEP 6 "Open the kimono." This unfortunate phrase is still commonly used in business circles to refer to the moment when a seller opens the books for a potential buyer to inspect. If you're satisfied by a buyer's indicative offer, this is the next step. In

a large transaction, the seller may set up a secure data room with no outside lines, where buyer agents can come, review confidential records, and leave empty handed. Cheap virtual data room services are also available but aren't as secure.

STEP 7 Cooperate with the buyer's due diligence. Now the buyer must perform an exhaustive audit of all your company's records in order to satisfy the need to understand what she's buying, right down to the last paper clip and pending lawsuit. This will likely take at least a month and involve significant expense on the buyer's part, which is why it's common to sign an agreement granting *exclusivity*—a promise that the company will not be sold to anyone else until a decision is made. There is a variation called *vendor's due diligence*, in which the seller contracts a trusted outside agency to conduct the audit.

STEP 8 Sign a contract. This is the final, formal, legally binding document that will be notarized and signed by both parties. This contract may run to dozens or hundreds of pages and must be prepared by a lawyer experienced in corporate sales transactions. Normally the buyer prepares the first draft, but you must hire an experienced lawyer to advise you on it. Problems at the contract stage should be worked out between the lawyers.

223 ASK FOR AN NDA UPFRONT

At any point in the invention process, you may want to have partners sign a *nondisclosure agreement* (NDA). But by this stage, when you're sharing secrets with potential buyers, it is an absolute necessity. No ifs, ands, or buts.

An NDA is a legal contract between two or more parties stating that the intellectual property, trade secrets, or anything else discussed in a business relationship cannot be shared or used for profit. For inventors, disclosing details about your concept with manufacturers, investors, and licensing companies is simply unavoidable. You can protect your IP (see #96–111), but the only thing stopping money-hungry go-getters from making a pile of dough off your idea is, well, you.

It's best to customize your NDA with a lawyer to make sure you've covered all likely scenarios. And save yourself a load of trouble (and maybe a confidentiality breach) by prescreening buyers—if someone doesn't have the capital to buy you, he doesn't need to see your bank statements.

224
MEET LONNIE JOHNSON
INVENTOR OF THE SUPER SOAKER

Over the past 25 years, the Super Soaker has become nearly synonymous with fun. For many of a certain generation, the sound of its pump sliding back and forth immediately conjures up a hot summer's day . . . mere moments before a powerful blast of water delivers a head-to-toe dowsing. Despite its reputation as a zany plaything, the top-selling "Wetter is better!" technology originated in the bathroom of Lonnie Johnson, an Air Force and former NASA engineer who was simply trying to build a more environmentally friendly heat pump.

At the age of 33, Johnson had already had a hand in several impressive engineering feats. He'd worked on the stealth bomber program and helped integrate a nuclear power source into *Galileo*, the spacecraft that eventually went to Jupiter in 1989. But Johnson was always tinkering at home in his off hours, and he was hard at work on developing a heat pump that used water instead of Freon when he made the discovery that changed childhood forever. To test a prototype late one night in his bathroom, he attached the device's vinyl tube to the water tap and aimed its nozzle into the tub. The resulting spray of water was so forceful it caught him off guard. He said to himself, "Jeez, this would make a great water gun." He then used a small lathe and milling machine to make the plastic parts for a prototype, which he gave to his seven-year-old daughter for a round of market testing. The neighborhood quickly reached consensus: His water gun blew all the others away.

Johnson's breakthrough discovery occurred in 1982, and on October 14, 1983, he applied for a patent, claiming that his invention incorporated an oft-used technology (a nozzle that sprayed high-pressure water) in a novel way (a squirt gun for kids). He originally wanted to manufacture it himself but—after receiving quotes of US$200,000

continued on next page

for the first production run of 10,000 units—he figured it would be better to find a licensing partner.

But by 1989, Johnson still hadn't been able to get his idea manufactured and out into the marketplace. Luck finally struck at the American International Toy Fair in New York City, where Johnson received an invitation to present his idea to Larami Corporation, a toy company located in Philadelphia—so long as he didn't "make a special trip." But that's exactly what Johnson did, armed with a new and improved prototype made of a 2-L bottle, Plexiglas, and PVC. "I opened my suitcase, took out the gun, and shot it across the conference room. And they said, 'Wow!'", Lonnie relayed in a piece he wrote for the BBC.

The rest, as they say, is history. Johnson licensed the water gun to Larami and worked with their team to revise the design so it could be sold for US$10. After a false start under the less-than-thrilling moniker the Power Drencher, the team switched to the more pleasantly alliterative Super Soaker and, with the help of a big TV ad push, sold 20 million guns in the summer of 1991 alone. It topped a staggering US$200 million in sales.

Over the years, Johnson continued to iterate on the Super Soaker, improving the pumping mechanism and widening the water stream to make the splash experience more enjoyable for the splashee. He even adapted the gun to accommodate Nerf projectiles after toy giant Hasbro acquired Larami in 1995. But Johnson also resumed his more academic work, creating film-thin rechargeable batteries and developing renewable energy solutions in the worlds of solar power and ocean thermal energy conversion. While most of Johnson's inventions have enjoyed success, it's also worth noting that some of his commercial products flopped; namely, a diaper designed to play nursery rhymes when soiled. (Hey, even the greats can't be great all the time.)

When faced with an initial disappointment at not being able to fund your own product, then another caused by a near-decade-long delay in seeing it successfully licensed, it's wise to remember Johnson, his estimated net worth of US$360 million, and some crucial words from the man himself: "I've never been very good at giving up."

FAQ

BORN
October 6, 1949, in Mobile, Alabama.

EARLIEST TINKERING
Johnson reverse-engineered his sister's doll in order to understand how its eyes opened and closed. He also built a go-cart using a lawnmower engine and scraps from the junkyard, and nearly burned down his parents' home making rocket fuel.

HIGH-SCHOOL NICKNAME
The Professor.

FIRST INNOVATION
In 1968, when Johnson was in high school, he was the only black student represented in the Alabama science fair at the University of Alabama at Tuscaloosa, where—just five years earlier—Governor George Wallace had tried to prevent two black students from enrolling at the school by standing in the doorway of the auditorium. Johnson took home first prize for his remote-controlled, compressed-air-powered robot, which he named Linux.

INSPIRING INVENTOR
George Washington Carver, a 19th-century inventor and botanist known for his products involving peanuts and sweet potatoes.

EDUCATION
B.S. in Mechanical Engineering and M.S. in Nuclear Engineering from Tuskegee University in Alabama.

CURRENT PROJECT
The Johnson Thermo-Electrochemical Converter System (JTEC), an incredibly efficient engine that turns heat into electrical energy via the compression and expansion of hydrogen gas.

With a maximum spray of 35 feet (10 m) and a pressurized blast that thoroughly drenches victims, Johnson's Super Soaker put all water guns that came before it to shame—and it stands the test of time too. Shown here in the iconic Super Soaker 50 version, here's how it won the famous water gun wars of the '90s.

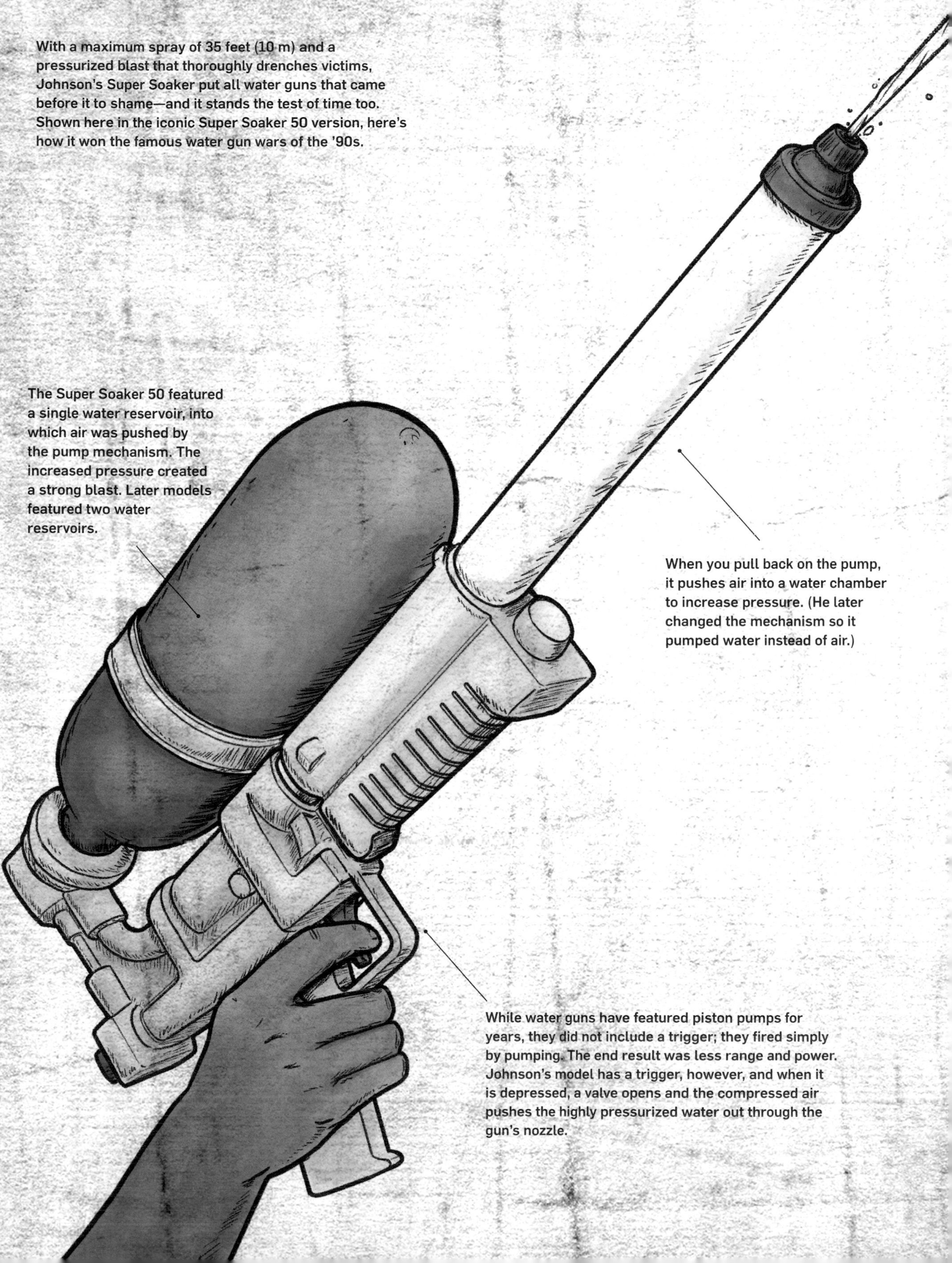

The Super Soaker 50 featured a single water reservoir, into which air was pushed by the pump mechanism. The increased pressure created a strong blast. Later models featured two water reservoirs.

When you pull back on the pump, it pushes air into a water chamber to increase pressure. (He later changed the mechanism so it pumped water instead of air.)

While water guns have featured piston pumps for years, they did not include a trigger; they fired simply by pumping. The end result was less range and power. Johnson's model has a trigger, however, and when it is depressed, a valve opens and the compressed air pushes the highly pressurized water out through the gun's nozzle.

INDEX

Disclaimer The information in this book is presented for an adult audience. While every piece of advice in this book has been fact-checked and field-tested where possible, much of this information is situation-dependent. The publisher assumes no responsibility for any errors or omissions and makes no warranty, express or implied, that the information included in this book is appropriate for every individual, situation, or purpose. Before attempting any activity outlined in these pages, make sure you are aware of your own limitations and have adequately researched all applicable risks. This book is not intended to replace professional advice from experts in business, programming, engineering, electronics, woodworking, metalworking, or any other field. Always follow all manufacturers' instructions when using the equipment featured in this book. If the manufacturer of your equipment does not recommend use of the equipment in the fashion depicted in these pages, you should comply with the manufacturer's recommendations. You assume the risk and full responsibility for all of your actions, and the publishers will not be held responsible for any loss or damage of any sort—whether consequential, incidental, special, or otherwise—that may result from the information presented here. Otherwise, have fun, and best of luck with your invention!

weldon**owen**

President & Publisher Roger Shaw
SVP, Sales & Marketing Amy Kaneko
Finance & Operations Director Philip Paulick

Senior Editor Lucie Parker
Project Editor Goli Mohammadi
Editorial Assistant Molly O'Neil Stewart

Creative Director Kelly Booth
Art Director Lorraine Rath
Senior Production Designer
Rachel Lopez Metzger

Production Director Chris Hemesath
Associate Production Director
Michelle Duggan

Weldon Owen is a division of Bonnier Publishing USA.

1045 Sansome Street, #100, San Francisco, CA 94111
www.weldonowen.com

Library of Congress Cataloging in Publication data is
available.

ISBN-13: 978-1-68188-158-4

10 9 8 7 6 5 4 3 2 1
2017 2018 2019 2020 2021

Printed and bound in Canada.

Popular Science and Weldon Owen are divisions of

BONNIER

POPULAR SCIENCE

Popular Science is the world's largest science and technology magazine, with 6.1 million print readers and 10 million monthly page views on PopSci.com. The publication explores the intersection of science and everyday life, providing science and tech news along with tons of fascinating DIY projects for beginning tinkerers and pro builders alike. Founded in 1872, *Popular Science* is one of the oldest continuously published magazines in the United States, and is published in five languages and nine countries.

PRODUCED IN CONJUNCTION WITH CAMERON + COMPANY
Publisher Chris Gruener
Creative Director Iain R. Morris
Senior Designer Suzi Hutsell

ILLUSTRATION CREDITS
Font cover and all section opener illustrations by **Joe Alterio**.

Author profile and all inventor profile illustrations by **Tim McDonagh**.

All other illustrations by **Conor Buckley**.

ADDITIONAL TEXT CREDITS
Profiles of Steven Sasson, Limor Fried, Helen Greiner, Peter Homer, Julio Palmaz, Bre Pettis, Ayah Bdeir, and Massimo Banzi by **Rachel Nuwer**.

Additional thanks to Steve Hoefer for writing #075.

PUBLISHER ACKNOWLEDGEMENTS
Weldon Owen would like to thank Katharine Moore, Marisa Solís, and Kevin Broccoli of BIM Creatives for editorial assistance; Sophie Bushwick at *Popular Science* for her brand consultation and content review; Rachel Nuwer for assistance with concept development; and all the inventors who were kind enough to share their stories with us.

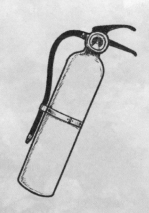

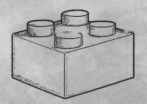